MORE THAN GENES

MORE

than

GENES

WHAT SCIENCE CAN TELL US ABOUT
TOXIC CHEMICALS, DEVELOPMENT, AND
THE RISK TO OUR CHILDREN

DAN AGIN

OXFORD
UNIVERSITY PRESS

2010

OXFORD

UNIVERSITY PRESS

Oxford University Press, Inc., publishes works that further
Oxford University's objective of excellence
in research, scholarship, and education.

Oxford   New York
Auckland   Cape Town   Dar es Salaam   Hong Kong   Karachi
Kuala Lumpur   Madrid   Melbourne   Mexico City   Nairobi
New Delhi   Shanghai   Taipei   Toronto

With offices in
Argentina   Austria   Brazil   Chile   Czech Republic   France   Greece
Guatemala   Hungary   Italy   Japan   Poland   Portugal   Singapore
South Korea   Switzerland   Thailand   Turkey   Ukraine   Vietnam

Published by Oxford University Press, Inc.
198 Madison Avenue, New York, New York 10016
www.oup.com

Oxford is a registered trademark of Oxford University Press

Library of Congress Cataloging-in-Publication Data
Agin, D. P.
More than genes : what science can tell us about toxic chemicals, development, and the
risk to our children / Dan Agin.
p. ; cm.
ISBN 978-0-19-538150-4 (alk. paper)
1. Prenatal influences.   2. Fetus—Abnormalities—Etiology. I. Title.
[DNLM: 1. Embryonic and Fetal Development—drug effects.
2. Embryonic and Fetal Development—genetics.   3. Maternal Exposure—adverse
effects.   4. Prenatal Exposure Delayed Effects—chemically induced.   5. Prenatal
Injuries—chemically induced. WQ 210.5 A267m 2010]
RG627.5.A55 2010
618.2′4—dc22   2009018102

1 3 5 7 9 8 6 4 2

Printed in the United States of America
on acid-free paper

*To the toxicologists and pediatricians who guard
the future of our species*

ACKNOWLEDGMENTS

I thank my wife Joan for doing no more than roll her eyes when I pounded the walls, Jeremy Katz for reminding me about reading science and drinking sea water, my agent Faith Hamlin for her grace and encouragement, and my editors at Oxford University Press, Marion Osmun and Catharine Carlin, for their hard work. Above all, I thank my long-gone father, unread, unfortunate, who bought his child a microscope with a whole week's wages.

CONTENTS

PART I
OF SHAPING AND DESTRUCTION

ONE

The Richness of Our Ignorance *3*

TWO

Pollution Babies *25*

THREE

From One Cell to a Hundred
Trillion *43*

FOUR

The Fetal Brain *65*

FIVE

Life In Utero: Shaping or
Destruction? *85*

PART II
THE FORGE OF CIRCUMSTANCE

SIX

The Endless Fetal Hangover *125*

SEVEN

Unborn Days and Sexuality *143*

EIGHT

DEVELOPMENTAL BRAIN DISABILITIES *169*

NINE

GENES, THE WOMB, AND MENTAL
ILLNESS *207*

TEN

MUCH ADO ABOUT IQ *233*

PART III
MISERY FOR ALL SEASONS

ELEVEN

CULTURE, POVERTY, AND FETAL
DESTRUCTION *263*

NOTES *309*

REFERENCES *329*

INDEX *379*

PART I

OF SHAPING AND DESTRUCTION

ONE

THE RICHNESS OF OUR IGNORANCE

During the next few decades, Americans and others in the industrialized world will learn that the psychological destinies of their children are often shaped and mangled by man-made environmental effects that begin not with birth but with conception. This is an idea that has been quietly gaining momentum in science for some years now, occasionally leaking into the popular press. As it becomes increasingly established, it will challenge the very fundamentals that govern the way we see ourselves and our society.

How will we deal with these effects? Are they real or mere speculation? When and how do they happen?

The origins are beyond what most people imagine.

On the morning of September 11, 2001, some 3000 people died in front of our eyes in a crazy scene of airliners crashing into skyscrapers and of those skyscrapers crumbling within minutes. Anyone downtown in Manhattan that day, or anyone anywhere in front of a television screen who watched the collapse of the twin towers of the World Trade Center, has the memory of it seared into their psyche. The entire appalling event—from Manhattan to the Pentagon to a small field in Shanksville, Pennsylvania—sent political shockwaves across America and around the world that have not yet subsided. We now call that day "9/11" as a signature shorthand for the catastrophe, a logo for an event whose details quickly occupied the mind of nearly everyone on the planet.

But like many catastrophic events, there was more to 9/11 than most people realize.

Not long afterward, a few miles north of "Ground Zero"—the empty ground where the World Trade Center once stood—a pediatrics group at the Mt. Sinai School of Medicine, together with others there and at the Bronx Veterans Affairs Medical Center, began to ask a simple question: Was it possible the shock of the 9/11 catastrophe had caused effects in the fetuses of pregnant women who lived close to the disaster?

The Mt. Sinai research team went on a hunt for pregnant women who had been in the vicinity of the World Trade Center at 9 a.m. on the 11th of September 2001. They published advertisements in local newspapers. They distributed flyers in lower Manhattan. They sent letters to 3000 obstetricians in the greater

New York City area. They found 187 women who had been pregnant and present in any one of five exposure zones around Ground Zero, including 12 women who were in the towers at the time of the attack. As a comparison group, they used 174 pregnant women who had been nowhere near the World Trade Center on the morning of the catastrophe.

The researchers analyzed every piece of relevant information available about the pregnant women in both groups and about the infants born to them in subsequent weeks or months. On August 6, 2003, they published a short letter in a medical journal. The concluding paragraph of the letter had no ambiguities

> We found an apparent association between maternal exposure to the World Trade Center disaster and intra-uterine growth retardation, suggesting that this event had a detrimental impact on exposed pregnancies . . . Possible long-term effects on infant development are unclear and will require continuing follow-up.[1]

Two years later, the Mt. Sinai research group published three papers on their findings in three different medical journals.[2] To sum up their conclusions: The cause of intrauterine growth retardation in the infants was apparently not dust and smoke inhaled by the pregnant mothers, but maternal psychological stress and cortisol secretion effects, as indicated by measures of below-normal cortisol levels in their infants.

The findings of the Mt. Sinai research group are not isolated. Since the late 1990s, fetal effects have been found from earthquakes, ice storms, and floods, with varying later outcomes for the children: childhood verbal deficits, depression, schizophrenia, and so on.

Do we know the mechanism for these effects? There's more than one possibility, but consider the following: On 9/11, when a pregnant woman was close enough to experience the traumatic World Trade Center event, her adrenal glands secreted the powerful stress hormone cortisol. Her cortisol entered the placenta. Not all of her cortisol was broken down by the placenta, and some of it got through to the fetus and increased the fetal blood cortisol level. Recent studies in fact show a positive correlation between maternal and amniotic fluid cortisol levels.[3] On 9/11, to compensate for increased local cortisol, the fetal adrenals reduced their own cortisol secretion to keep the total level down. But since that happened while the fetus itself was developing, the result was fetal production of cortisol that might not have been just transiently reduced, but permanently reduced. One effect could be retarded intrauterine growth. Another effect would be low cortisol levels in infancy (as found by the Mt. Sinai group) and later consequences difficult to assess. In other words, during development the fetus adapted to a new environmental condition as if that condition would be permanent.

In modern pediatrics and developmental psychobiology, this adaptation is called "fetal programming" or "prenatal programming." It's a new concept. The general idea is that during development important physiological parameters can be reset by environmental events—and the resetting can endure into adulthood and even affect the following generation—in this case, producing a transgenerational nongenetic stress disorder.

So what are the consequences? Researchers have already correlated heart disease and diabetes with prenatal growth and apparent fetal programming.[4] But "intra-uterine growth" is only what you can see and determine by measuring an infant's head circumference and body length at the time of birth. What you can't see are the subtle effects on various physiological systems, for example, on the developing central nervous system—on the developing brain. You can measure behavior later on, but it's not that easy.

What is certainly true is that you don't need great drama—earthquakes or floods or terrorism—to affect the prenatal environment. Far subtler events can have an impact on that environment as well.

SPERM, EGG, AND THE STUFF OF HUMAN DEVELOPMENT

We tend to forget the richness of our ignorance—how much we don't know and how recently we learned what we do know. When, for example, did we know about the intricacies of human reproduction, and how did we come to know them?

Only 330 years ago, the janitor of the Delft City Hall in Holland was an amateur scientist with an obsession for grinding lenses. He was 45-years-old. He was also the first human being to describe human sperm cells. But when Anton van Leeuwenhoek announced his observations by correspondence to the British Royal Society, he did so with much reticence. He was fearful that what he wrote to the Society would be thought obscene.

And sure enough, van Leeuwenhoek's work spread ripples of consternation throughout the Western civilized world. We already knew that the mating of a human male and human female produces children, and so it seemed obvious that male "seed" had something special to do with it. But tiny swimming animals? What did one do with this sort of observation of reality? Shout explanations, of course. Maybe the loudest shout was the idea that if these swimming animals did indeed have something to do with reproduction, then each swimming "animalcule" must carry inside its tiny self a complete individual in a tiny form—an homunculus. How else to explain the transmission of human qualities from one generation to the next? People shook their heads in wonder at this new evidence of the intricate marvel of human existence.

Thus we discovered male sperm cells before we knew anything about female egg cells. Another 163 years passed before we had the notion that the signal event in human reproduction involves penetration of the female egg cell by the male sperm cell. By that time the belief that a tiny individual sits inside the front end of each sperm cell had been discarded, and we understood that sperm and eggs contained something—stuff, molecules, chemicals, something—and that the joining of the something in the sperm cell and the something in the egg cell sets off a process, a cascade of events inside the mother. Determined by what? We had no idea. And how were the human qualities transmitted from one generation to the next? It was certainly true that children tended to resemble their parents. But why are we born humans and not tigers? Again, we had no clue.

In 1859, an English gentleman-bird-watcher, who had done some traveling in far away places, published a detailed account of his observations and conclusions about why certain qualities are transmitted from one generation to the next, how and why there are so many varieties of living things, and how higher forms are derived from lower forms. He wrote

> From the war of nature, from famine and death, the most exalted object which we are capable of conceiving, namely the production of the higher animals, directly follows. There is grandeur in this view of life, with its several powers having been originally breathed into a few forms or into one, whilst this planet has gone cycling on according to the fixed law of gravity. From so simple a beginning endless forms most beautiful and most wonderful have been, and are being, evolved.[5]

The English gentleman's name was Charles Darwin, and his ideas, coupled with the ideas of many dozens of observers before and after him, gave us the bedrock view of what we are and where we've come from.

Of course, bedrock views derived from science are not always welcomed by the public and institutions of power. Embryologists faced derision, mockery, and Church ostracism for their views about human reproduction and fetal development. As for Darwin, maybe no scientist after Galileo has been vilified as much. Even scientists, including biologists, derided Darwin. Louis Agassiz, Professor of Geology and Zoology at Harvard University, called Darwin's theory of evolution a mania and said he hoped to outlive it. Adam Sedgwick, the eminent British geologist who was one of the founders of modern geology, sneered that when he read Darwin's *On the Origin of Species* he laughed until his sides were sore. Humans and apes with a common ancestor? Never! Pages could be filled with the rants of Darwin's contemporaries who spit, fumed, and shouted that he was mad as a hatter. (Many hatters of the time were indeed crazy: they used mercury compounds to make felt cloth for hats, and the mercury poisoned their brains.)

America, a country first colonized by religious fundamentalists, always hated Darwin. In the 1920s, two generations after Darwin published *On the Origin of Species*, Americans were still frothing at the mouth about his theory of evolution, their fever maybe best expressed by William Jennings Bryan, who announced with his chin out front

> All the ills from which America suffers can be traced back to the teaching of evolution. It would be better to destroy every other book ever written and save just the first three verses of Genesis.[6]

But science goes its own way. During the 1930s, the combining stuff of sperm cells and egg cells was recognized to involve genetic entities (genes) that were parts of a very large molecule—deoxyribonucleic acid, or DNA. After the Second World War, in the early 1950s, a reasonable structure for that molecule was proposed by Francis Crick and James D. Watson, and the complete package of DNA, genes, and evolution changed forever the way we think of ourselves.

So the "DNA Age" began, launching an era of extraordinary research in molecular biology and genetics. Before long we had attempts to explain human behavior in terms of genetically determined hard-wiring of the human brain and the evolution of individual human traits and social behavior—that is, the emergence of a new field called "sociobiology," a presumptive new evolutionary psychology of man.[7] We've now had more than 50 years of the DNA age, and as might be expected, early enthusiasm caused more than a few ideas to be oversold.

To avoid confusion, here's my brief:

I am a scientist—a psychologist and a biologist. To me, the theory of evolution is on ground as firm as the theory that the Earth revolves around the Sun and that the Sun is a star like other stars. What is also on firm ground is the idea that the shuffling and natural selection of groups of genes and gene regulators are the major mechanics of biological evolution. But what is not on firm ground, not yet, are the detailed biological, environmental, and cultural determinants of human behavior. Those determinants, despite what sociobiologists may have to say about them, are far from settled.

Modern science is a dynamic enterprise, and major changes in scientific thinking often occur while the public is still adjusting to the last change. There is, for example, new thinking that natural selection can operate on more than one level, not only at the level of individuals, and that the rush in the 1960s and 1970s to discard group selection (natural selection of groups) was an error.[8] There is also new thinking that the old nature versus nurture controversy was badly framed, that environment can seriously affect the expression (regulation) of genes—that is, determine which genes are switched on and which genes are switched off—and that in humans the prenatal environment can be a major determinant of gene regulation and later behavior.[9] And, finally, there has been a shift of attitude among psychologists, biologists, and anthropologists about the details of the origins of human behavior, particularly about the effects of the prenatal environment.

The development of a human being from a fertilized ovum has been compared to a lump of iron somehow turning itself into the space shuttle. Wondrous, yes. But evolution is what produced the mind-boggling program of human prenatal development—it did not drop down from a cloud in the sky. We're still far from understanding all the details of the process, but we'll get there eventually. Meanwhile every biologist on the planet who knows anything about cells and tissues and organisms is now fully aware that the nine months of human prenatal development constitute the most vulnerable and formative period in the life of any individual.

THE EMERGENCE OF NEW REALITIES

What's the science that directs us to consider the prenatal environment so important in childhood and adult behavior? The evidence and conclusions have been accumulating for more than a decade and form the substance of this book. Some highlights:

By the early 1990s, it was already obvious to researchers that ethyl alcohol in the prenatal environment can cause prenatal damage. Measurable concentrations of maternal blood alcohol produce the recognizable clinical entity of fetal alcohol spectrum disorder (FASD). But on biological grounds, even minute immeasurable quantities of ethyl alcohol are expected to cause subtle effects on the developing nervous system, especially during the earlier stages of embryonic formation. These effects might involve damage other than FASD or its most severe form, fetal alcohol syndrome (FAS)—subtle damage not yet clinically recognized. There is, in other words, no apparent safety threshold for ethyl alcohol consumption, and no apparent safe period after conception—which poses a dramatic problem for women who drink alcohol before they even know they're pregnant. Of course, like every other toxic effect, individual susceptibilities vary. But the potential for damage to the fetus is always present.

In 2002, a news article in the journal *Science* highlighted low birth weight as a predictor of adult chronic disease, the article pointing out that epidemiological studies of such correlations have been accumulating since the 1980s.[10] Birth weight is merely a crude measure of the rate of fetal development. Are there subtle behavioral effects of lower than average birth weight that have not yet been tested in children and adults? We already have some evidence that birth weight is correlated with childhood and adult intelligence quotient (IQ) and various behavior patterns.

In 2003, researchers reported a study of 48,197 mothers and their 59,397 children.[11] The children were followed until age 7, and the group included 623 twin pairs. The report concluded that in poor families nearly all the variance in IQ is accounted for by a combination of effects in both the prenatal and post-natal environments, and that the contribution of genes to IQ variance in such families is close to zero. In contrast, in affluent families, the result is almost the reverse. The implication is that among poor children genetic differences contribute almost nothing to the measured variance of IQ because environmental damage, both prenatal and postnatal, overwhelms all other variables in accounting for IQ variation. In contrast, in the middle and upper classes, in which prenatal and postnatal damage to the nervous system is much reduced and hardly variable from one family to the next, genetic differences account for most of the variation in IQ. In plain language, for middle and upper class children, differences in IQ can be explained mostly by genetic differences, while in lower class children, differences in IQ are explained mostly by nongenetic differences (prenatal and postnatal environments).

In 2004, a review of the impact of environmental toxins on childhood health pointed out that in Europe the prenatal environment is at risk of contamination by 15,000 synthetic chemicals, nearly all developed over the past 50 years, and that worldwide between 50,000 and 100,000 chemicals are being produced commercially.[12] The review emphasized that end-organ vulnerability may be greatest in the fetus, and that for some chemicals the toxic effects seen in childhood are the result of continuous exposure beginning with conception.

In 2004, a news article in the *Journal of the American Medical Association* (*JAMA*) affirmed that the prenatal environment may have profound long-term consequences for health. For example, stress during pregnancy (as I've indicated) can create problems for the fetus. Evidence shows that a woman who is stressed during her pregnancy is at greater risk of bearing a child who will develop an anxiety disorder or behavioral problems. The *JAMA* article called for more detailed research on multigenerational data to determine the long-term effects of the prenatal environment on individuals.[13]

In 2005, a review pointed out that exposure to infection or nutritional deprivation during early fetal development may elevate the risk of later onset of schizophrenia.[14] Lead exposure has also been established as a risk factor for schizophrenia, the lead correlations suggesting environmental chemicals as a new class of schizophrenia risk factors that are still inadequately investigated.[15]

In 2007, a news article in the journal *Science* discussed maternal effects on schizophrenia risk, the article pointing out that not widely appreciated in deducing the importance of genes from twin studies is the fact that two-thirds of monozygotic (identical) twins share a placenta, which is a key environmental factor. Individual placentas vary with respect to the transport of various nutrients and hormones, which affects normal development. X-chromosome inactivation is affected by placental status, and in the largest study of its kind, so is IQ. The article emphasizes the possibility that the placental environment can influence the expression of genes linked to neurodevelopment and schizophrenia.[16]

In February 2008, 200 researchers who had met the year before in Norway announced that embryos and fetuses are more vulnerable than previously thought to chemical pollutants that can cause disease or disability, and that embryos and fetuses are vulnerable even to extremely small doses that do not harm adults.[17]

These and a host of other studies since the early 1990s, research to be discussed in this book, all point to a new attitude in biological and medical science—a new understanding that the prenatal environment is an extremely important determinant of health, disease, IQ, and behavior.

THE SCHEME OF THE BOOK

This book has eleven chapters in three parts. In Part I, we deal with the reality that each of us experiences two environments—prenatal and postnatal—during our lifetime. The first environment is probably at least as important as genes and the postnatal environment in determining the genesis of childhood and adult behavior. This part includes, in the next chapter, a detailed examination of the effects of chemical pollution, specifically lead, to provide a broad framework for understanding how an environmental toxin that is all around us can, in often subtle ways, affect the prenatal development of the brain and nervous system. The biology of fetal neural development is then explored in the remaining chapters of Part I, addressing such questions as, Why is the fetus so vulnerable to damage? How does evolution produce and constrain human embryonic development? Is the mother an enemy or ally of the fetus? What in the prenatal environment are the changing and even destructive influences—toxic and otherwise—on brain development that can affect later childhood and adult intelligence and behavior?

In Part II, we consider some of the postnatal costs of those influences and examine the strong evidence that the prenatal environment is related to various behavioral dysfunctions or anomalies and to deficits in intelligence. The chapters in this part cover fetal alcohol disorders, transsexuality, developmental disabilities such as attention deficit hyperactivity disorder (ADHD), autism, and psychosis, as well as the emotional frameworks that give rise to a few other mental disorders. The important question of individual IQ is also addressed, including the controversies about its measurement in IQ tests (what are such tests? Do their scores signify anything? How is their reputed significance used?) and the extent to which the prenatal environment comes into play in the variance of IQ.

In Part III, we move past the individual costs of prenatal disruptions to address broader social and economic questions that relate to those costs. What, for example, are the effects of ethnic and cultural heritage, geographic location, and socioeconomic status on the prenatal environment? The chapter in this part also seeks to answer questions about the relation between prenatal impacts and social behavior such as criminality.

My intention throughout the book is to make the case that the prenatal environment must be considered in any attempt to understand the origins of human health and disease, and of human behavior and intelligence, and in any resolution of the simplistic nature—nurture debate put to the public by the media. Nature is more than genetics, and nurture is a more complex proposition than what occurs in just the postnatal environment of parents and communities and nations. "Environment" starts at conception, and neglecting this fact as we have for too long has resulted in often tragic consequences for those born into our world and, by extension, for us all.

Some people may complain that much of the evidence presented in this book suggests rather than demonstrates the neurological effects of negative impact, in or outside the womb, on the prenatal environment. That caveat is worthy and clear, but it's also clear that when a smoke alarm goes off in your house, suggesting a fire, the most prudent action is to find out what's happening. The evidence I present sounds that alarm, and in so doing, it helps frame an understanding both of what we know and of what we don't yet know about the nature of that fire, in this case the pernicious effects of environmental assaults to the unborn. The evidence we have now is thus as much a call for future research as it is, to shift metaphors here, a dead canary in a mineshaft.

A few words about the type of evidence I present. For every study involving human subjects or patients, there are ten studies involving animals, and for every study involving animals there exists the question of how applicable its results are to human biology and medicine—especially when animal behavior must be measured. It's not easy to measure the emotions or the cognitive performance of a rat or a rhesus monkey. Animal behavior, even chimpanzee behavior, is not merely simplified human behavior.[18] Sometimes the judgment of applicability is easy, but often it's extremely difficult and maybe outside the interest of the general reader. In this book, then, I've tried as much as possible to focus on the results of studies with human subjects and patients. Some places in the book are replete with numerical data about the prevalence of clinical diagnoses or of particular behaviors or about other topics relevant to these concerns. I do understand that for many people reading such paragraphs is like drinking sea water. But the remedy is to sip slowly, one sentence at a time. The numbers can be revealing, and it's better to have them than to avoid them—better to know, for example, that one clinical disorder is more prevalent than another clinical disorder and that prevalence can depend on what city or county you live in.

Also, in public health, quantitative data often change from time to time and from place to place. There is nothing static about public health or human society in general. Hence, wherever possible, I've avoided controversies about numerical values by choosing a single value that I think is probably representative of a given phenomenon.

Presenting one particular set of numbers here allows me to emphasize a few other points about the content of this book. A significant number of live newborns in America come into the world with structural defects—which means defects in organization of tissues in brain, organs, bone, and so on. Some statistics[19]: Major structural anomalies (for example, cardiac defects) occur in 2–3 percent of live-born children, and an additional 2–3 percent are recognized by age 5 years. In approximately 50 percent of cases of birth defects the cause is unknown. Of the known fraction, genetic factors are believed to account for 15 percent, environmental factors for 10 percent, a combination of genetic and environmental factors 20 percent, and the effects of twinning producing maybe as much as 1 percent. Minor anomalies (for example, minor visual problems) occur in approximately 15 percent of newborns.

At the present time, the best we can say is that about half the recognized fetal anomalies are caused either solely or partly by impact on the fetal environment. Here "partly" means the anomaly evidently involves some combination of genetics and environment. I'm not concerned in this book with gross malformations of the fetus—the branch of biomedicine called teratology (dysmorphology)—but with changes in prenatal development due to impact on gene expression that can cause later changes in behavior and intelligence in the child and adult. The main organ system addressed in this book is thus the brain, and since the most important feature of the brain is its internal organization, many changes important for its function are invisible to the naked eye—no visible malformations at all. In fact, it often happens that two brains can be quite different in behavior—extremely different—without any visible differences at the level of gross histology (visible tissue). So whether we can observe a gross tissue change in the brain doesn't tell us whether a change in the brain's neuronal circuitry exists.

IF IT'S PRENATAL ENVIRONMENT, WHO IS TO BLAME?

I look at many possible causes of fetal damage in this book, but give particular emphasis to man-made environmental toxins, be they chemical pollutants in our air, water, and land, or substances in the consumer marketplace, legal and otherwise. The effects of these toxins are often connected to more than one type of fetal damage. Prenatal exposure to lead, for example, is connected to postnatal occurrences of lower IQ, ADHD, and schizophrenia, usually in different individuals. Prenatal exposure to tobacco smoke, as another example, is connected to a host of effects with adverse postnatal consequences as well. The effects of both sorts of toxins can also be related to other fetal impacts, which in turn are often connected to important social consequences such as criminality. Who specifically is to blame for these and other man-made effects? Who is responsible for this fetal damage and the consequences that come of it?

Pregnant mothers are responsible for smoking cigarettes, using cocaine, drinking alcohol, and so on, but their behavior is subject to enormous cultural pressures and pressures of circumstance. "Responsibility" for personal behavior is complicated. For example, an educated woman living in Rome may be aghast that rural Italian pregnant women drink wine steadily during pregnancy, but if the local culture holds that wine is good for the fetus, the rural pregnant woman is merely doing what everyone around her is doing. What does "responsibility" mean in such a case? Such questions are relevant around the world, and I attempt to address them in various places in this book.

But certainly pregnant mothers are not responsible for lead in the air or in the paint on toys, or for arsenic in the water, mercury in fish, polychlorinated biphenyls (PCBs) everywhere, unaffordable adequate nutrition, unaffordable preventive health care, or the daily stresses of surviving poverty. Indeed, most of the effects that damage the unborn have absolutely nothing to do with personal responsibility. For the most part the responsibility belongs to society writ large—to local, state, and federal government, to industry, to an unknowing or misled public, and to media ballyhoo that ignores science and instead puffs heredity as the primary determinant of human behavior and intelligence.

America currently has a heavy bundle of social, economic, and political problems related to the prenatal environment, but

the burden of responsibility for allowing these problems to continue unchecked is one that as citizens we all must share. What
we can responsibly do about them—what, for example, we can
do to bring about change in our public health policies to safe-
guard the fetus—is begin by learning what science can tell us
about the prenatal environment and about the possible causes
and consequences of fetal damage.

A TEACHER AND HIS RINGDOVES

Nearly 60 years ago I walked into an introductory psychology
course taught by a young man named Daniel S. Lehrman. He
reminded me of what was at that time a new television personality—a rotund fellow named Jackie Gleason. Twice a week for
2½ hours each session, Lehrman glided back and forth in front
of the class, often talking about birds, particularly ringdoves.
He had us all spell-bound with his reductions of "instincts" to
environment, experience, physiology, and biochemistry.
Lehrman's lectures were among the last of their kind. A few years
later, James D. Watson and Francis Crick announced their proposed structure for DNA and biology changed forever. Hardly
anyone talks of "instincts" anymore, reduced or otherwise.[20]
Now it's genes and heredity and genetic origins of behavior—
and some of it is a media carnival.[21]

Fifty or sixty years ago, there was already a feeling among
psychologists and neurobiologists, led by people like Lehrman,

that the concept of instincts was too simplistic, not enough in it to explain the complexities of human behavior. Something similar is happening now as we realize the idea of genetic "hardwiring" of behavior is also too simplistic. So the shuffle begins: new ideas, new experiments, more new ideas. That's how science works. We move forward, peering into the shadows of our ignorance. For over 30 years now, too many people have been bemused by overly simplistic ideas about the origins of human behavior. It's time to move beyond them and to take a closer look at the possible origins in our earliest hours and days and months.

TWO

POLLUTION BABIES

During the summer of 2007, Americans learned that they had a big problem with imported toys for children. Paint containing lead was long ago banned for use in consumer products in the United States, but the toy industry apparently depends on toys made in China. Riddled with lead, those toys were sold by the millions in the United States before the contamination was discovered. A panic set in, and for good reason: lead is a serious neurotoxin, toxic for adults, for children, and especially for fetuses. During fetal development, for example, chronic exposure to low levels of environmental lead is correlated with low birth weight and loss of hearing. It can also cause birth defects, mental retardation and other distinct IQ deficits, autism, reading disabilities, attention deficits, and a constellation of other learning problems in school. Lead exposure can also sometimes be fatal.

We can give this picture a human face. Meet Roberto, a young boy living in a major city in the eastern United States.[1] Aged four and a half, he's neatly dressed, alert, seems comfortable in the room, with no signs of any emotional stress. But it doesn't take long for you to understand that something is wrong: he never engages you in conversation.

If you ask, "How are you, Roberto?" his answer is, "How are you, Roberto?"

If you say, "The weather's nice today," his response is, "The weather's nice today."

Does he understand what you say? He avoids eye contact and his facial expression never changes. But if he sees a movement, he'll quickly turn his head and follow it. A slight movement of someone's hand will catch his attention and he'll focus on the movement immediately. If a marble rolls on the table in front of him, he turns immediately and watches the marble.

You say: "Did you have cereal for breakfast this morning?"

His face impassive, he replies: "Did you have cereal for breakfast this morning?"

Roberto has an autistic disorder characterized by marked impairments in communication and other social interactions, and by restricted interests and activities.

Is his condition due to a genetic glitch? Or did he have a prenatal trauma that interfered with the development of his brain? Or have there been emotional problems in his interactions with his parents and siblings? What we know is that the concentration of lead in his blood is five times the national average for children his age—and that previously, when he was 2 years old, it was 20 times the national average. He already showed symptoms of developmental deficits at that earlier age. He's been receiving anti-lead treatment (chelation therapy) ever since. Unfortunately, the treatment only reduces the lead in his blood. There's no treatment that will remove the lead already in his brain.

Fast forward a few years, and Roberto is now 6 years and 10 months old. His blood lead level (BLL) is still several times the national average for children his age. Some of his autistic symptoms have disappeared, but he still has great difficulty in social interactions. Now he responds to simple commands, but in conversation he still repeats back what you say to him (echolalia). His verbal communication is so impaired, it's almost impossible to test his cognitive function.

At this age, Roberto's brain is irreversibly damaged. He's attending a school for developmentally disabled children. He's not suffering from classical autism, but will likely be living for the rest of his life with various cognitive disabilities that will remain a severe handicap.

"THE NATIONAL AVERAGE"

If Roberto's BLL had been at or even under the national average, would he have been safe from the effects of lead toxicity?

Here begins a slippery slope of changing criteria for determining the threshold at which one is or is not considered lead poisoned.

The current median BLL for U.S. children 1 to 5 years old is 2 micrograms per deciliter (2 units). In 1980, the median was 15 units, although distributions appeared drastically skewed: In a survey conducted between 1976 and 1980, 98 percent of American black children between the ages of 1 and 5 years had BLLs equal to or greater than 10 units; the corresponding figure for American non-Hispanic white children was 84 percent.[2] Meanwhile, the U.S. Centers for Disease Control (CDC) has invoked various BLL standards over the years as a "threshold for concern"—the current "threshold" of lead poisoning. In 1965, the CDC standard was 60 units. In 1975, it was 30 units. In 1985, it was 25 units. Its current BLL standard of 10 units was instituted in 1991 (according to which most American children in the 1976–1980 survey would now be said to have been experiencing the effects of lead poisoning), but within a decade it was apparent from clinical research that even the 10-unit standard was too high. Local averages are often even higher.

Indeed, there's mounting evidence that even "low levels" of lead exposure during the fetal and early postnatal periods are still great enough to cause important cognitive deficits, if not severe disability.[3] A review of data from Boston; Cincinnati; Cleveland; Rochester; Mexico City; Port Pirie, Australia; and Yugoslavia shows that these effects have been confirmed in different locations and with different populations.[4]

So if your local city or county average is 10 units (or 10 micrograms per deciliter), and your child has a BLL of 10 units, it's likely your child already has an intelligence quotient (IQ) deficit of 7.4 points compared to a child with a BLL of only 1 unit. If your child has a BLL of 60 units, it's likely that the IQ deficit will be 31 points. That's enough to cause a child who would be of average intelligence to be mentally retarded.

And the child with a BLL of only 1 unit isn't "safe" either. In fact, as far as we can tell, there's no threshold of safety for lead in the blood, no concentration below which lead is harmless to the child's developing brain. The only safe concentration is zero. That's the consensus of clinical researchers and pediatricians in America and around the world.

Meanwhile, the numbers are grim. In 2001, according to the U.S. Environmental Protection Agency (EPA), approximately 11 percent of the children in Chicago under 6 years of age had enough lead in their blood to be considered lead poisoned.[5] The same EPA report asserted that by 2007 at least 10 million American children below the age of 6 years had measured levels of lead in their blood high enough to impair the development of their brains.[6] Another estimate, published in 2004, suggested that during the past 15 years two out of three children in some communities (e.g., Oakland (California) and Chicago) have been lead poisoned by their environment. The figure for Los Angeles is 32 percent lead poisoned. In the year 2000, approximately 22,000 children under 5 years of age in Wisconsin could be classified as lead poisoned.[7] According to the CDC, between

1999 and 2002, 310,000 children in the United States aged 1 to 5 years had BLLs that classified them as lead poisoned.[8] Before that, from 1991 through 1994, approximately 900,000 children aged 6 years and younger had blood lead concentrations in the lead-poisoning range.[9] By 2007, in some urban areas in Massachusetts, as many as 22 percent of all children were lead poisoned.[10]

Worldwide, all developed countries have lead-pollution problems similar to those of the United States.[11] In Australia, for example, a blood lead survey in 1996 found that 7.3 percent of preschoolers had levels greater than 10 units—certainly enough to cause brain damage. In 2004, a European survey reported some countries with as much as 30 percent of the population with BLL greater than 10 units.

As for developing countries, according to the World Health Organization, 120 million people around the world have BLLs greater than 10 units and 240 million people have levels greater than 5 units. Among children, 78 percent in South Africa had levels greater than 10 units in 2002. In Jamaica, 42 percent of rural children and 71 percent of urban children had BLLs greater than 10 units in 2000. In India, 51 percent of urban children had BLLs greater than 10 units, and 13 percent had levels greater than 20 units, in 1999. In China, 34 percent of children had levels greater than 10 units in 2004.

Of all these children worldwide, how many might have been exposed to lead prenatally? This is hard to say, but it's reasonable to assume that high BLLs in infants or very young children, like Roberto, would likely have been high in utero as well. When a pregnant woman is exposed to lead, it crosses into the placenta, bringing about a blood lead concentration in the fetus that is similar to that of the mother,[12] who acts as an intermediary between the fetus and the maternal environment. We can measure the prenatal BLLs of both mother and fetus by analyzing her blood and the lead in the umbilical cord blood. After that we can analyze the BLLs in children at various ages. But lead in the body does not all remain in blood—it gets sequestered into bone and brain, where measurement is impractical or impossible. Thus the BLL is probably only a minimal reading of the lead that's actually in the body as a whole.

In any case, it's estimated that the human fetus is ten to a hundred times more sensitive to ambient lead than children or adults, such that the so-called national averages are almost certainly dangerous for the fetus. In 2006, researchers reported a study of 146 pregnant women in Mexico City.[13] It's one of the few studies to measure maternal lead values during each trimester of pregnancy. The researchers examined the impact of prenatal lead exposure on fetal neurodevelopment by measuring whole blood and plasma levels of lead in the pregnant mothers at each trimester and then in umbilical cord blood at delivery. When the infants were at 12 and 24 months of age, the researchers measured their BLL and also evaluated their neural development with a standard method (the Spanish version of the Bayley Scales of Infant Development). From the evidence, the report concluded that prenatal lead exposure has an adverse effect on neurodevelopment, and the effect is most pronounced when exposure occurs during the first trimester of pregnancy. The report suggested that screening and intervention any later than the first trimester may be too late to prevent the greatest damage to the fetal brain and nervous system from lead pollution. The average maternal BLL in this study during the first trimester of pregnancy was 7 units. This value for the United States is lower but still dangerous.

LEAD: OUR ENDURING NEMESIS

The plain fact is that in any modern society, industrialized or becoming industrialized, lead is everywhere. It's in the air, the soil, the water, and our food. Some of the lead in our environment has origins across many centuries. It's estimated that the processing of lead ores during the past 5,000 years has released 300 million tons of lead into the environment—most of it within the past 500 years.[14]

Leaded gasoline has long been banned in the United States and some other countries, but not before millions of tons of lead poured into the air for 60 years from the exhausts of trucks and automobiles. The lead from those exhausts settled on everything around us, particularly in our soil. Lead has also entered soil from underground leaded gas storage tanks, industrial operations that produce lead grindings, and lead paint waste.

Indeed, lead-based paint is now the major source of lead pollution in America. The "toy scare" is only a small part of our problem. Lead was used extensively as a corrosion inhibitor and pigment in both interior and exterior oil-based paints before 1978, and some paints were manufactured with lead concentrations of 50 percent by weight. Most buildings older than 60 years, residential or commercial, have lead-based paint somewhere inside or outside of them. Even if the paint has been carefully scraped away, its dust has to settle somewhere inside or close to the building. The weathering of lead-based exterior paint and deposits of paint chips and dust remain a significant source of lead in soils surrounding homes, with soil lead concentrations at or above 500 micrograms per gram carrying particular risk: a pregnant mother living near that soil and breathing the dust it emits will have a blood lead concentration that equals or exceeds 10 units.[15]

Another possible source of ingested lead is drinking water from household plumbing fixtures, including metal pipes, faucets, and soldered joints. The lower the pH of the water (lower pH = higher acidity) and the lower the concentration of dissolved salts in the water, the greater is the solubility of lead in the water and the more easily is lead leached from pipes carrying the water. Leaching of lead from plastic pipes has also been documented and may be due to the use of lead stearate, a stabilizer used in the manufacture of polyvinyl plastics.

Other common sources of polluting lead are small-arms munitions—bullets and shot—and fishing sinkers. Lead is often used as a coloring element in ceramic glazes. Some candle manufacturers, particularly outside Europe and North America, may treat candle wicks with lead to make them burn longer; zinc is an alternative, but it's more expensive. The cosmetic kohl, imported from the Middle East, India, Pakistan, and Africa, usually contains lead. Folk remedies and herbals, particularly Ayurvedic medicines, are also often contaminated with lead.

The key point here is that common lead is a stable chemical element, and like all stable nonradioactive chemical elements, it's indestructible. Once it's mined and refined and allowed to pollute the environment through gasoline engine exhausts, smelter emissions, paint dust, and the like, it's with us forever. Lead is now the most widely scattered toxic metal on earth, its toxicity unreduced by the passage of time. Some other equally unsettling facts are as follows [16]:

1. Owing to the short half-life of lead in the blood (only several weeks—after that it moves into tissues and bone), the period during which lead can be detected in the blood can be far shorter than the duration of its toxic actions in the brain.

2. Once deposited in the brain, lead is eliminated extremely slowly—a brain-half-life of approximately 2 years.

3. Once in the brain, lead cannot be removed by chemical chelating agents of the sort that Roberto received. This means that even after the lead levels in blood have been decreased by chelation therapy to apparently insignificant

concentrations, the lead that has already been deposited in the brain continues its neurotoxic action. Can the effects of lead on the developing brain eventually be reversed? There's not enough known yet to answer the question for all cases. We do know that chemotherapy of children like Roberto, with blood levels above approximately 20 units at first measurement, usually does not produce any complete reversal of cognitive injury. The disconnect between blood levels of lead and brain levels of lead complicates the outcome of treatment.

4. The consequence is that once an elevated blood lead concentration has been detected, it's too late to prevent lead-caused brain damage.

5. If we combine all of the above with the recent strong evidence that even very low BLLs can seriously affect brain development and subsequent cognitive function, we are left with a succinct policy conclusion: the only way to prevent fetal and childhood lead poisoning is to prevent lead from ever getting into the bodies of pregnant mothers and their children in the first place.

Of importance is that some hereditarian psychometricians are fond of measuring behavioral and IQ differences between various "racial" and ethnic groups and then offering us conclusions about how genes determine behavior and IQ.[17] Do such psychometric studies of heritability of behavior and IQ involve adequate controls for prenatal environment? They do not.

In fact, measured differences between groups *do* depend substantially on differences in prenatal environment. In America, the populations of pregnant women (and their fetuses) who are most exposed to environmental lead pollution are those living in the inner cities. They tend to live in old buildings containing leaded paint or dust from the scrapings of old leaded paint, and the soil around their buildings is contaminated with lead. These old buildings also tend to have lead pipes in their plumbing systems. We know from measurements that these women and their children have high levels of lead in their blood.

For example, studies show that 50–70 percent of children living in the inner cities of New Orleans and Philadelphia have BLLs above the current EPA guideline of 10 units. In contrast, in Manhattan, where very little soil is exposed, only 5–7 percent of children have levels greater than 10 units. Just across the East River in Brooklyn, where exposed soil is common, the levels in children are several times higher. In general, the lead concentration in soil of old communities in large cities is 10–100 times greater than comparable neighborhoods of smaller cities.[18]

Although we now recognize that BLLs below 10 units can be dangerous to the developing brain, in the 1940s a level of 40 units was considered ordinary and safe, and pediatricians were reassured by the apparent absence of symptoms of lead poisoning in their patients.[19] But absence of symptoms is not evidence of absence of damage. And sometimes that damage can be quite subtle, owing in part to the limitations in our diagnostic tools for assessing intellectual and especially behavioral deficits.

How much of what we see as differences in mental abilities are due not to genes but to differences in environmental exposures of fetuses and children to lead and other neurotoxins? Researchers estimate that at the extremes of IQ there could be a three-fold increase in retardation, and a loss of two-thirds of the extremely gifted. Other research links lead to behavioral school problems of children independent of effects on IQ.[20] Such behavioral effects—which are complex, not easily quantified, and often considered merely variations in behavior among individuals—are difficult to assess, but many studies confirm the correlation.

Let's consider just a few of those effects in brief detail here. In the brain, lead probably corrupts more than one biochemical system. It interferes with events at nerve cell connections called "glutamate synapses," which means at approximately half the synapses (connections) in the cerebral cortex. These synapses are apparently critical for learning during development. The important glutamate receptor that's involved is the so-called NMDA (N-methyl-D-aspartate) receptor. This receptor is selectively blocked by lead, and the behavioral consequence is a deficit in learning ability.

Recent evidence from functional magnetic resonance imaging (fMRI) studies of children followed from birth to adolescence suggest that childhood lead exposure (and by implication, fetal lead exposure) may have a significant and persistent impact on the ongoing reorganization of brain connections involved in language function.[21]

High levels of lead in the body are associated with juvenile delinquency and crime,[22] and some researchers believe behavioral problems ascribed to attention deficit hyperactivity disorder (ADHD) may be due to low-level lead exposure. I will discuss these topics in greater detail later in the book.

Problems in lead-induced NMDA receptor function have also been proposed as a cause of schizophrenia. Between 1959 and 1966, 12,000 pregnant women gave samples of blood serum while they were pregnant. The blood samples were frozen and stored for later analysis. Thirty-eight years later, in 2004, researchers analyzed the samples of 44 women whose children subsequently developed schizophrenia and 75 other women in the group whose children did not.[23] The research team reported that offspring of mothers whose blood lead concentration was greater than 15 units were twice as likely to develop schizophrenia as the offspring of mothers whose BLL was less than 15 units. We don't know yet whether the schizophrenia-lead relationship will be supported by more extensive studies. If the link is confirmed, the scope of the lead pollution tragedy will increase enormously.

It's important to note that not all developing embryos and fetuses have the same vulnerability to environmental lead. Given two individuals with the same blood lead concentration, it's possible that one individual will show clear symptoms of neurological damage while the other individual will show no clinical symptoms at all.[24] We don't know why these differences in vulnerability exist or what the mechanisms are that determine vulnerability. We just know that the vulnerability exists and that even at low levels of lead in the blood, heme synthesis (critical for the function of red blood cells) and other biochemical systems are affected, potentially resulting in psychological and neurobehavioral impairments that for countless children can be delayed and substantial.

Lead pollution is everywhere, and as the earlier statistics indicate, it is poisoning children by the millions. This is not a "natural" calamity. Lead in deep earth, before mining, is immobilized in rock in combination with other elements and it's not harmful to us. It became harmful only when we mined it, refined it, smelted it, produced toxic lead compounds and refined elemental lead for commercial use, and then effectively sprayed millions of tons of toxic lead everywhere.

We made this calamity ourselves—and we're still making it.[25]

BEYOND LEAD

I have considered lead pollution in detail, but lead is not the only villain in the prenatal environment. Others include methylmercury; polychlorinated biphenyls (PCBs); dioxins; pesticides; ionizing radiation; and maternal use of alcohol, tobacco, marijuana, and cocaine. These villains can cause a range of behavioral effects from severe mental retardation and disability to subtle changes in mental function that depend on the timing and dose of the chemical agent. Indeed, more than 200 industrial

environmental agents have been demonstrated to have neurotoxic effects in humans—with probable critical windows of vulnerability for the fetus—and the neurodevelopmental disorders
caused by these industrial chemicals may constitute a "silent
pandemic" in modern society.[26]

In 1999, researchers reported a study of fetal outcome following maternal occupational exposure to common organic solvents
in occupations, such as factory workers, laboratory technicians,
artists, office workers, and so forth. The data showed that occupational exposure to common organic solvents during pregnancy is associated with a 10-fold increased risk of major fetal
malformations.[27] This is occupational exposure, not some dramatic accidental acute poisoning. By implication, subtle effects
on the fetal nervous system, effects that shape emotional and
intellectual behavior, are to be expected after maternal exposure
to ambient concentrations of organic solvents ordinarily used
in many workplaces.

The brain and nervous system are unique tissues, with different parts responsible for different functions, the parts developing at different times. For example, the brain neural circuits
for motor control, sensory response, intelligence, and attention
do not develop at the same time, and each development has its
own window of vulnerability during the life of the fetus from
conception to birth. Also, many cell types in the brain have
different windows of vulnerability with varying sensitivities to
environmental agents.

What research on the prenatal effects of pollution shows us is that we need to be extremely careful about drawing conclusions based on uncontrolled psychometric studies of individual and group differences in behavior and IQ. Research on prenatal effects refutes the very idea that the impact of the environment on the fetus is of no consequence in assessing the origins of human behavior and intelligence, and the origins of differences between groups. As later chapters in this book will demonstrate, the new evidence of dramatic effects of environmental pollutants on neurodevelopment in the fetus tells us that the prenatal environment constitutes an important set of variables determining who we are.

THREE

FROM ONE CELL TO A HUNDRED TRILLION

The year 1978 was not a happy one for many people. The American inflation rate was more than 12 percent, the unemployment rate was more than 6 percent, and the Cold War dragged on without apparent end.

The year did have its ups, however, and one of them was heralded on the 12th of August by a brief and unassuming letter that appeared in the British medical journal *The Lancet*. "Sir," it began, "we wish to report that one of our patients, a 30-year-old nulliparous married woman, was safely delivered by caesarean section on July 25, 1978, of a normal healthy infant girl weighing 2700 grams."

Nulliparous means the woman had not given birth to any previous children. The authors of the letter were Patrick C. Steptoe, an obstetrician and gynecologist, and Robert G. Edwards, a biologist and physiologist.[1]

Now why would such a routine obstetrical event appear in one of the most respected medical journals in the world? The headline over the letter gave the answer: "Birth After the Reimplantation of a Human Embryo."

Nine months before, on the 10th of November 1977, an egg cell had been removed from one of the ovaries of a 30-year-old woman named Lesley Brown. She had blocked fallopian tubes, which prevented any egg cells released by her ovaries from becoming fertilized by sperm cells. Lesley Brown's removed egg cell was instead fertilized "in vitro," literally "in glass," in this case in a flat petri dish. Two and a half days later, when the fertilized egg cell had divided three times to produce a clump of eight cells, the clump (called a morula) was implanted into Lesley Brown's uterus to begin the usual 9-month gestation.

The infant girl that showed up the following July was named Louise Joy Brown. The procedure that produced her, in vitro fertilization (IVF), has since produced millions of new people around the world. Louise herself is now in her thirties, a smiling, happily married woman with a child of her own (conceived in the ordinary way).

We've come a long way since the seventeenth century, when Anton van Leeuwenhoek first looked at sperm cells with one of his home-made microscopes. But the achievement of Steptoe and Edwards was not easy. Many others had experimented with IVF before and had failed to produce successful results. Indeed, this repeated failure to achieve IVF before the success of Steptoe and Edwards was already a signal that the importance of the prenatal environment for development is not something to be waved away with a flick of one's hand.

THE NOT-SO-EASY PATH TO BIRTH

The human ovum is the largest cell in the body, on average 145 microns in diameter (average human hair width is 100 microns, or 0.1 millimeters), which means it's visible to the naked eye. The ovum is about 15 times larger than ordinary cells, such as skin cells and liver cells, but it's still no larger than a dot, smaller than the period at the end of this sentence. The profound glory of human reproduction, the wonder of wonders, is that under the right circumstances over about 277 days of gestation, this biological dot is capable of turning itself into a 7-pound infant ready to scream at you to look smart and give it some food and attention.

The turning of a biological dot into an infant human being is a process. For millions of years we knew nothing about the process or about how to manage and protect the newborn it created, and most infants died of one cause or another either during or immediately after birth. These days, most infants live, many maturing into adults with a life span of maybe 80 years in advanced countries and 45 years or less in undeveloped countries. Nonetheless, these survival rates, even in the worst places on Earth, are usually much greater than the survival probability of a fertilized human ovum: Egg cells, blastocysts (early embryos at the time of implantation), and later embryos and fetuses are extremely vulnerable to environmental effects, the program of gestation easily diverted or shut down completely.

More than 50 percent of all fertilized human egg cells (ova) never complete development, never finish the process that turns a biological dot into an infant human being, never result in a live birth. The curtain usually closes before the mother is even aware that she has conceived. There are so many circumstances that can terminate prenatal development; it's also a wonder of wonders that the process ever gets completed at all.

But even the circumstances leading to a live birth are reflective of the vulnerability to change in prenatal development. Identical (monozygotic) twins, for example, share the same genome, but as individuals (phenotypes), they are often not similar at all, their differences a result of differing environmental effects producing differing gene expression—and the differences can arise early in embryonic and fetal development.[2]

GENETICS 101

We'll return to the issue of prenatal vulnerabilities—their whys and wherefores—but first we need to discuss what surely is everyone's favorite topic: genetics. *Genomes* and *genes* are big buzzwords these days, with some people in the media tossing them around willy-nilly as though they know what they mean.

Not long ago, a well-known pundit announced on TV with his usual gravitas that embryonic stem cells are unique because they carry the entire human genome. But the fact is that every cell in the human body, except certain cells of the immune system, carries a copy of the entire human genome (actually two slightly different copies). The cells on the tip of your nose carry the entire human genome.

The term "genome" refers to all the genes in the human body, plus intervening nongene DNA. Also called "junk DNA," nongene DNA is now considered to be less junk than before since the discovery that it codes for special ribonucleic acid (RNA) involved in one way or another in regulating genes.[3]

Genes themselves are often described as if they're laid out like the beads of a necklace, with various lengths of nongene DNA separating the genes. But this picture is not correct. In the first place, the necklace is not a single strand; it exists in pieces (chromosomes). And in the second place, and most important, we now know that genes are not unitary entities.[4] Parts of them can be in different places and even in different chromosomes. When a gene is active, its transcription (production of a corresponding RNA molecule) involves splicing together transcribed parts from various places. The more we learn about transcription, the more the term "gene" loses its precise meaning.[5]

And this is very relevant for the way we think about fetal development. The more complex the process that "reads out" genetic information stored in the genome, the more complex is the vulnerability of that process to environmental impact.

Think of the Internet: As its complexity has grown during the past few decades, so has the magnitude of its vulnerability to interference, shutdowns, scams, mishaps, and so on.

But we need to be cautious with our analogy. In gestation, not every interaction is disruptive. Many interactions of the fetal environment and local cellular environments with genes are important triggers during development: triggers for tissue growth, for cell migration, for hormone secretion, and so on.

This idea needs emphasis. As the embryo and fetus develop in the womb, the developing entity is continually interacting with its local environment, just as a child or adult interacts with his or her local environment. There is nothing "automatic" about the development of any embryo, and that includes the human embryo and its later form, the human fetus.

Of course, some of the old ideas remain intact. Specific DNA code and RNA product, whether unitary or transcribed and spliced from various places, ultimately governs the synthesis of specific proteins, the large molecules that play either a structural or mechanical role or act as enzymes to catalyze specific chemical reactions or act as gene regulators of one kind or another. The entire set of chromosome DNAs, genes plus junk, is the genome. Genes code for proteins, but we now know that some genes can code for more than one protein (depending on how their RNA-product pieces are spliced together), and some proteins can be coded for by more than one gene.

Maybe by the end of this century the term "gene" will be considered archaic. Meanwhile, the term is still here, although continuing research makes its meaning fluid. We should not be surprised. Nature doesn't care much about how we use our language to organize what we see. The handicap of the structure of our language is our own burden, not the burden of nature.

GENE EXPRESSION

What makes a liver cell different from a brain cell? Although they both carry the same genome, carry all the genes and all the intervening junk, not all genes in any particular cell are active, and those that are active in a liver cell are a different set than those that are active in a brain cell. At any one time, certain genes are active, switched on, "expressed," their codes used in the synthesis of specific proteins. The set of expressed genes in a liver cell differs from the set of expressed genes in a brain cell.

In this book, I will use the term "gene expression" to refer to the ultimate production of a protein or RNA molecule derived from the information stored in DNA. Gene expression can be blocked, changed, or regulated, and its changes apparently can occur in three main ways: (1) in the DNA molecule itself (for example, DNA methylation); (2) in the way DNA is folded and arranged to form the chromosomes (for example, histone modification); and (3) biochemically anywhere between the read-out of DNA and the synthesis of the ultimate product determined by that piece of DNA code (for example, RNA interference).

Chromosomes are made of packaged chromatin (nucleo-histone), which in turn consists of DNA and proteins. Most of the chromatin proteins are of a special type called histones. Chromatin was once considered a static thread of packaged DNA and histones, but it's now understood to be a dynamic structure involved in the regulation of many aspects of DNA biochemistry and gene expression. Specific proteins access and read the DNA sequences in chromatin, either to transcribe these sequences or to mediate gene expression. But for this to occur, chromatin must be locally unraveled to expose DNA sequences. Biochemical processes that unravel chromatin are therefore crucial for transcription. Any environmental event during fetal development that affects the unraveling of chromatin can block or unblock transcription or change gene expression.[6]

But at this point I want to focus only on a single important general idea: it's the genome that stores the inherited biochemical information of our species, but it's gene expression that determines how and when that information is used, and gene expression itself is regulated both by internal gene regulation networks and by interaction with the environment.

There is, then, no such thing as a simple genetic substrate (a genotype) imparting "destiny" to an individual biological system (a phenotype). Development (ontogenesis), which begins with the fertilization of the ovum, is a result of complex interactions between the genome and the environment, and that is what needs to be the focus of any understanding of human behavior.

Certainly, to understand the details of fetal development, the idea of varying gene expression is absolutely necessary. Given a genome, various types of cells have different sets of genes in the genome activated or not activated, switched on or off, by various mechanisms. The most important thing we've learned in the last decade or so is that gene regulation almost never occurs without the influence of nongene entities, events, and environments. This process—the interaction and effects of nongene variables with and on gene regulation—is called "epigenetics." It is itself the focus of an exploding new field that overlaps molecular, developmental, and evolutionary biology and that now produces mores than 2500 articles a year in scientific journals. Definitions of "epigenetics" vary.[7] Some people use the term in a narrow sense to describe gene–environment interactions that reprogram development, but in this book I will use it in its broad sense to indicate any gene–environment interaction that changes gene expression, or more specifically, to define any change in gene expression caused by an interaction with the local cellular or global organismal environment.

For example, if a pregnant mother gets emotionally stressed and her adrenal glands secrete cortisol, that hormone travels through her blood, into and through the placenta, and can affect gene expression in the developing embryo or fetus, pushing that development one way or the other, or even wrecking it. This is an epigenetic effect caused by an interaction between the maternal environment and fetal gene expression. In another example, epigenetic changes in gene expression in a critical region of the brain (the hippocampus) were recently found on autopsy in adults who committed suicide and who had a history of child abuse and severe neglect.[8]

Again, the essential point is that the genes and junk DNA that form the genome do not determine the destiny of the individual. Individual destiny is determined by gene expression, which is always influenced by interactions with and triggers by the environment. No organism, including humans, has a destiny independent of its environment.

DETERMINANTS OF HUMAN
BEHAVIOR AND INTELLIGENCE

Our basic terminology about genes and gene expression in place, we can now consider the three sets of variables that determine human behavior and intelligence:

(1) The first set of variables lies in the genome. These variables are the fixed genes in each individual and the changing and changeable repertoire of actually expressed genes (whose differential expression is triggered by local interactions) that program the development and early hard-wiring of the nervous system.

(2) The second set of variables lies in the prenatal environment. These are variables that the mother either produces (her own hormones, for example) or filters from external sources (tobacco-smoke chemicals, for example) into the placenta and the local environments of prenatal tissues. During gestation,

many of these variables affect various gene expressions involved in the development of the brain and nervous system. We will call them "epigenetic events"—effects that result from interactions with the environment that change gene expression.

(3) The third set of variables lies in the postnatal environment. These include the physical environment (outdoors and in) and the social environment (family, peers, neighborhood, socioeconomic status, education, culture, and so on). Many of these variables affect the development of emotions, learning, language skills, and the processing of information by the brain—and can also affect gene expression via stress, infection, malnutrition, and toxins in air (for example, lead), water (for example, arsenic), and food (for example, mercury). Those effects that occur during early development can be extremely critical.

Most discussions of the origins of human behavior and intelligence focus only on comparing variables in groups (1) and (3), neglecting the importance of the prenatal environment, the variables in group (2). But in fact, all three groups, including the interactions among the variables in them, play a role in those origins. No human being develops from conception through gestation to birth, then through infancy, childhood, adolescence, and into adulthood, without some influence from these various environments. Throughout the life span, the human brain and nervous system continue to develop, with neural connections made and unmade. This is especially so in childhood, during which the brain and nervous system change dramatically in response to numerous factors, including the effects of experience.

In fact, our major gift from biological evolution is a brain whose intimate structure can change through experience, a brain thoroughly adaptable after birth. That adaptability, called "plasticity" by neuroscientists, coupled with a language that passes learning from one generation to the next with high efficiency (especially when it's a written language), is the major watershed separating human evolution from the evolution of other animals. As a result of our extensive neurological ability to learn from experience, most of our individual behavior is learned, and because of the transgenerational passage of learning, most of our group behavior is also learned.

The public has been pushed into the somber shadow of belief that its destiny is fixed in concrete by inherited genes, that groups differ in behavior and intelligence because of their inherited genes, and that the Victorian idea that "Whatever is, is right," should be our guide in public policy. But our new knowledge of the wonders of DNA, exciting and important as it is, doesn't tell us even half the story of why we humans behave the way we do. Nor does it factor in the extent to which that behavior, and the workings of the brain and nervous system that give rise to it, is affected by what happens in the womb. Prenatal development is not a process we can dismiss, as those in the Group 1 and 3 camps, including those obsessed with the nature-nurture debate, would prefer to believe.

Nor, as we already know, is it an easy process; prenatal development is the most vulnerable period in the biological history of any human being. There are entire ranges of prenatal effects caused by circumstances known and not yet known, ranges that are not lethal or crippling but still of tremendous consequence for prenatal development and the health and behavior of the postnatal individual.

THE VULNERABILTY OF PRENATAL DEVELOPMENT

Which brings us to a critical question: Why is prenatal development so vulnerable to damage?

There's much that happens before fertilization that might affect prenatal development—for example, maternal clinical problems, or inherited genetic errors, or DNA damage by mutations or environmental toxins—but we'll focus here only on prenatal development from fertilization to birth. With the fusion of a sperm cell and an ovum at fertilization, the phase of the human life cycle of prenatal development starts the long and wondrous journey from a single cell (the ovum) to the trillions of specialized and organized cells of a new individual at birth. What we're looking at is a cascade of many thousands of events, a cascade with specific vulnerabilities at specific times—and the possibility for several vulnerabilities at any single time.

A cascade is a succession of sequentially interdependent events, each event both triggered by the event preceding it and itself acting as a trigger for the next event. But the cascade of prenatal development is more complicated. It involves not only multiple events occurring and interacting at any instant before the next set of multiple events is triggered, but also multiple interactions with the local environment. And these local interactions can themselves be necessary triggers in the cascade.

The ovum starts with a genome. Throughout development, as the number of cells derived from the ovum continues to increase, the genome in each cell is the same, but which genes are turned on (expressed) changes, producing specialized (differentiated) cells or new triggers for subsequent events—release or take-up of specific biochemical entities, rearrangements of cells into tissues and organs, migrations of cells to new locations in the developing embryo and fetus, appearance of specialized cell organelles, and so on.[9]

The cascade of development is essentially chemical, involving not just a few kinds of molecules but hundreds of thousands of kinds of molecules, organizing, rearranging, moving from one place to another, enhancing (catalyzing) the synthesis of themselves and other kinds of molecules. The kinds of molecules in the cascade are so numerous that we've hardly yet catalogued even a small fraction of them. Biochemists estimate that a single cell may contain as many as a hundred thousand different proteins, carbohydrates, and lipids, and no one anywhere has yet identified more than a small fraction of that number.

The cascade of prenatal development is thus a cascade of gene expression events, chemical events, and cellular events—and the whole mix moves forward by both internal triggers and triggers brought about by interactions with the local environment.

So the first important cause of prenatal vulnerability is complexity: the sheer complexity at many levels of prenatal development means that an enormous number of different and important process points are available for disruptive effects.

Another important cause of prenatal vulnerability is pace, the high rate of cellular proliferation necessary to transform a single microscopic cell (the fertilized ovum) into a 6- or 7-pound newborn infant consisting of trillions of cells specialized and arranged to constitute the human body externally and internally—albeit in the small of the infant. For example, it's estimated that in the developing brain and nervous system of the prenatal human, about 250,000 new neurons are generated each *minute* at the peak of cell proliferation during gestation.

The high rate of cell proliferation means a high rate of metabolism, chemical synthesis, cellular rearrangements and migrations, conversion of maternal nutrients into fetal cells and tissues, and so on. In prenatal development, everything is happening rapidly. If any specific process has its rate changed up or down by an unscheduled impact with the local environment, the consequence may be anything from a subtle bending of development in one direction or another to a lethal corruption that kills the embryo or fetus.

In general, the metabolism of embryos is different from that of adults, and the fetal construction of an organ can be affected by chemicals that have no apparent damaging effect on the normal adult functioning of that organ.

The third major cause of prenatal vulnerability involves size and simple physics. If a small permeable mass—a cluster of cells, for example—is exposed to a chemical, that chemical can reach all parts of the mass quickly by simple random diffusion. With larger masses, the diffusion time to reach all parts increases dramatically. But as late as the 6th week of gestation, the human embryo is still only a quarter of an inch in length and has no developed circulatory system, and any freely permeating chemical that gets into the embryo by any route will quickly diffuse throughout the embryo to affect every embryonic cell.

Throughout the embryonic period, until the 10th week of gestation, the situation is not much better. At the 10th week, when we begin to call the developing embryo a "fetus," we're dealing with an embryo/fetus only about 2 inches in length, indeed recognizable as a vaguely human form, but still small enough for simple diffusion to quickly distribute any permeating chemical entity throughout its body.

Small size facilitating distribution by simple diffusion is one of the reasons the early weeks of prenatal development are so vulnerable to certain chemical effects. The other important reason is that those early effects can be multiplied as the cascade of development proceeds. For example, there is mounting evidence that a critical window of vulnerability for fetal alcohol spectrum disorder occurs very early—during and shortly after the blastocyst stage—and that alcohol affects early gene expression in the developing embryo.[10] It appears that concentrations of alcohol too low to produce gross morphological disruptions may still cause subtle changes in the connections between nerve cells in the developing brain. Evidence indicates, as we'll see in a later chapter, that among the results of those changes are cognitive deficits in children whose mother drank alcohol during pregnancy.

What happens during prenatal development, then, is not a simple scheme but a set of schemes whose complexity demands the utmost respect. Are we merely guessing about all these schemes and interactions and vulnerabilities? Definitely not. During the past 100 years, mammalian experimental embryology has provided a mountain of evidence to shape our picture of prenatal development. If the entire period of prenatal development is thought of as a movie, all the individual frames of the film are not yet known, but we know the script, the cast, the plot, the major scenes, and we're working our way through the film step by step. We have reduced human prenatal development, the transformation of a biological dot to a 7-pound screaming infant, from a breathtaking mystery to a set of tractable scientific problems—be they problems in normal development or in the destructive environmental effects on development.[11]

EVOLUTION AND HUMAN PRENATAL DEVELOPMENT

Tucked away in the wilderness of central California is a mountain range called the Evolution Group. Its mountain peaks, named in 1895 during a federal geological survey, are called Mt. Mendel, Mt. Darwin, Mt. Huxley, Mt. Haeckel, and Mt. Wallace after some of the seminal figures in evolutionary science. To have a look at these peaks, you need gumption and perseverance. First you go to a town named Bishop about a dozen miles north of Death Valley. The town has a population of 3500 and it calls itself "The Mule Capital of the World." From Bishop, you drive west on Highway 168 to the edge of Lake Sabrina, get out, leave the car, and enter a national park wilderness trail to the Evolution Group. Mt. Darwin is the first and easiest to approach. North of it is Mt. Mendel and south is Mt. Haeckel.

Nearly everyone in the world knows the name Darwin, but outside of evolutionary biologists and people who like to read about the history of biology, probably the only people who know the name Haeckel are those living in a place called Jena in Germany, where Ernst Haeckel was born in 1834 and spent most of his life until he died in 1919.

Ernst Haeckel started out as a botanist, but was pushed into the study of medicine by his father. A year after receiving his medical degree, he decided he couldn't tolerate the misery of his patients and that he would rather be a zoologist. At the age of 25, he was appointed Professor of Comparative Anatomy at the University of Jena and he taught zoology there for 47 years. A mountain in California is named after him because he was Germany's Huxley, the biologist who promoted Charles Darwin and the theory of evolution to the German intellectual community and the German public.

As interesting, and the primary reason for this digression about mountain peaks, is the fact that Ernst Haeckel proposed a grand scheme about the development of embryos that was first hailed by biologists as magnificent, then in the next generation cast aside as ridiculous, and finally in still another generation, a few years ago, recognized as not completely ridiculous and maybe a key to some profound insights into the mechanics of evolution.

Despite all the controversy about Ernst Haeckel—his now infamous early writing on eugenics used posthumously by the Nazis to bolster their racist ideology, the rise and fall of his theory of development, the accusations that some of his drawings of embryos represented more wishful-thinking than reality—at least some of his scientific ideas, now in vogue again, are of outstanding importance.[12] For without question, Haeckel must be considered one of the ultimate founders of the new and exciting field of evolutionary developmental biology—the field that attempts to explain prenatal development in all its details. If we're to understand the importance of the human prenatal environment for later childhood and adult behavior, it is this field that will provide us with the hard ground of basic biological science.

Ernst Haeckel made errors in his observations, drawings, and conclusions about prenatal development, but the fundamentals he emphasized were correct: All vertebrates develop a similar body plan as embryos, an arrangement consisting of a notochord, body segments, pharyngeal pouches, and other structures. The commonality of the developmental program among vertebrates is a reflection of a shared evolutionary history. Even more significant, and this is confirmed by overwhelming evidence that researchers have gathered in recent years, it is clear that development in different animals is controlled by common genetic mechanisms, a watershed in our understanding of prenatal human development from conception to birth.[13]

Phenotypes (individuals) are expressions of genotypes (genomes), and it's often the case in the animal and plant world that for a single genotype there are various expressed phenotypes depending on the local environment. The idea that phenotype programming is always fixed by genotype and independent of environment has been obsolete for nearly a century, but unfortunately still promoted by some genetic determinists.

Yes, genes ultimately control development, but the control is not simplistic. These days the most important idea in developmental biology is that it's the switching of genes on and off (expression or prevention of expression) that determines the direction development takes at any instant in the embryo and fetus, and that the switching is heavily dependent on interactions with both local tissue environments and the global prenatal environment.

The human cascade of prenatal development is controlled by gene-switching. Most of the developmental genes involved are identical in humans, chimpanzees, and rats, but which switches are turned on or off, and where and when and in what order, is determined both by evolved regulating genes and by local and global cues many of which involve the prenatal environment. Thus, the prenatal "development" cascade is in fact a prenatal "gene-switching" cascade.

The development of a human fetus is a chemical manufacturing process that takes maternal oxygen derived from the air that the mother breathes, and maternal nutrients derived from the food that the mother eats, and by cascades of programmed biochemical-reaction routines turns the mix into various tissues and physiological organizations by switching genes on and off during the 9 months of gestation.

But there's more. What the new field of evolutionary developmental biology is telling us is that the gene-switching cascade evolves, that it's impacted, that it interacts with the environment, that it's subject to natural selection. In other words, whereas previously we thought of natural selection as acting only on reproducing adult phenotypes interacting with their environment, we must now conceive of natural selection as acting also on the prenatal developmental gene-switching cascade interacting with its environment. That process, too, is an important part of the mechanics of evolution. Indeed, the formation of whole species may itself be primarily a developmental phenomenon. The most ridiculous and stultifying idea in biology may be that it's possible to discuss genotypes and consequent phenotypes without also discussing the various ways in which epigenetics can bend, alter, and divert expression of phenotypes—with certain paths self-regenerating into new species.

In short, if we are ever to completely understand evolution, we must conceive of it in a way that incorporates a thorough understanding of prenatal development. No doubt in the future biologists will look back and wonder why it took us nearly

150 years to merge our ideas about development with those of evolutionary theory. Perhaps if they read the history of biology from Darwin forward, they will understand why: Getting to the place where such a unified vision is embraced has not been easy—with or without mountains named after evolutionary biologists.

FOUR

THE FETAL BRAIN

One way to excite undergraduate students in a neurobiology course is to carry in a glass canister containing a preserved whole human brain. You remove the brain from the canister and place it on a flat tray on the table at the front of the lecture room.

You tell them it's a whole human brain, and many mouths drop open. "Is it real?" they ask.

You tell them it's not a model; it's a real preserved whole human brain.

The usual next question is, "Whose brain is it?"

Since you have no idea, you tell them it's your mother's brain.

They wince. They laugh. They relax. Then you invite the entire class to form a line and come to the table one at a time to look closely and touch the brain if they want. So they do it, some with reluctance, some with fierce interest in their eyes; some only lightly touching the brain after protecting their hands with plastic, while others probe the longitudinal fissure and spread the hemispheres of the brain. No matter what their approach, all of them look bemused and even a bit mystified.

Now you go out and return to the lecture room with another brain, this one with its two hemispheres completely cut apart, two halves of a human brain now sitting on the demonstration table beside the whole brain. You ask the students to look at the inside. They gather around the table peering at the separated two halves. One of the students slips her hands into plastic gloves and lifts half a human brain in her two hands. You will see the wonder in her eyes.

It's the wonder in their eyes that makes your reward. Millions of years, you tell them. Millions of years of human evolution and here we stand examining a human brain and pondering its complexities.

The human brain confronted this way is a profound enigma. It's hard to imagine that our hopes, fears, sadness, happiness, amusements, passions, and all our acquired learning occur in a mass of tissue like these specimens on the demonstration table. This pulpy mass, this great tangled knot that defines our species, this transcendent puzzle reduces many other puzzles to trivialities.

And its development in a human embryo or fetus is perhaps the most transcendent puzzle of all. As with the development of other organ systems that occurs during 9 months of gestation, human brain development involves a cascade of processes in time. But it's not a single cascade, it's a branching tree (an arborization) of cascades with many events occurring at the same time. Think of brain development as enormously complex serialized chemistry along many interdependent biochemical reaction paths. At any particular time during gestation, we can expect not just a few, but many points of interaction with local environments—and many points of vulnerability to divergence or disruption of one or more ongoing cascades.

Those are the key words: divergence or disruption. Suppose that toxin molecules are present in the local environment of an ongoing cascade and that they interact with the surfaces or interiors of differentiating cells. In general, they can cause a developmental cascade to diverge in a new direction that may have minor or serious ultimate consequences, or they may disrupt a cascade altogether, stop it dead in its tracks, and prevent that branch of the development cascade-tree from growing any further.

Fetal brain development is a vast subject, the focus of hundreds of research laboratories and clinical groups around the world. In this chapter I highlight some of this research to illustrate how the fetal environment can interact with brain development and to identify broadly the points at which that development is most vulnerable to environmental effects that may alter it in some way, whether divergently or disruptively.

Subsequent chapters will explore in increasing detail some of the consequences of these effects.

BRAIN PLASTICITY

With an organ as complex as the human brain, many features of its structure and function can be called important, but maybe its most important attribute is its *plasticity*, a term neuroscientists use to identify the functional and anatomical malleability of the brain in response to experience. Of particular significance are neuronal connections in the cerebral cortex—that is, the plasticity of cortical synapses that apparently makes memory and learning and new responsive behavior possible.

The gross plasticity of the brain can be remarkable. Hemispherectomy, the removal of most or all of one hemisphere of the brain, is the most radical procedure in neurosurgery. It's used only as a last resort in the treatment of patients with certain brain cancers or with massive debilitating seizures. The surgery was originally performed only on the right hemisphere, because it was believed that since the left hemisphere controlled speech and language functions, removal of the left hemisphere would produce too much cognitive damage. Then, for many years, the technique was abandoned because of postoperative problems. When neurosurgeons revived the procedure, not only for the right hemisphere but also for the left hemisphere,[1] they demonstrated that in many cases language function recovers.

When asked how it's possible for patients with half a brain to live and have a life, the standard answer is "plasticity, flexibility, redundancy, and potential" of the human brain. But in reality, neurosurgeons don't know how it's possible. They just know that individuals can have half of their brains removed by surgery and be nurtured back into a useful functioning life. That's a fact that needs to be accounted for by any theory of how the mind works.

So we have a marker for understanding the brain: If, for clinical reasons, large parts of one hemisphere of a child's brain are removed by surgery, there may be no significant impact on that child's cognitive development.[2] Cognitive outcome depends more on the original pathology responsible for the surgery than on which or how much of one hemisphere is removed.

Not only is this an emphatic demonstration of the plasticity of the human brain cortex, but it also constitutes a great puzzle for those who consider cognitive abilities already hard-wired by genes at birth. Since it's not reasonable to assume that the brain cortex of a child suddenly becomes plastic after hemispherectomy, the reasonable assumption (and an assumption supported by much evidence) is that plasticity of the brain cortex is there at the start. The proverbial slate may not be completely blank in early childhood, but it's apparently blank enough to support the idea that there's much that will be written on that slate.

During the past decade the idea of human brain plasticity has been reinvigorated by evidence of neurogenesis (the production of new neurons) in adult primates,[3] and in many other adult mammals. There's also some evidence that the new neurons are usually interneurons in the brain cortex.[4] Interneurons are neurons with essentially short-range connections that connect other neurons whose interaction paths are much broader. In interconnecting larger input-output neurons, interneurons would be involved in plasticity underlying memory and learning. It's not possible to investigate neurogenesis in humans with the techniques that are used with animals, but maybe before long we will have firm evidence of neurogenesis in adult human cerebral cortex, and the propositions of hard-wiring enthusiasts will be even more problematic.

Meanwhile, the plasticity of the cerebral cortex, its ability to have its connections shaped by experience, is one of its outstanding properties, and understanding its mechanisms makes possible an understanding of what governs the wiring of the brain. It also raises the implication that any prenatal impact that affects plasticity in the fetal brain can seriously wreck the brain development that occurs both before and after birth.

BRAIN WIRING

Brain wiring, both during development and during childhood, is a complex process, and it seems that the rules governing it are variable from one anatomical place in the brain to another (and probably from one time to another). At short distances, local interactions between neighboring synapses can produce clustered plasticity, a malleability of connections of whole groups of neurons.[5] For example, the cortex of the human brain, only 2 to 4 millimeters in thickness, consists of six layers. It's a highly corrugated sheet because of the way the brain lobes are folded, and the connections of its approximately 10 billion nerve cells produce anything you think important that comes out of a person's head. The general architecture of the cerebral cortex is apparently programmed by triggered gene expression, but as development proceeds, local connections are more and more dependent on input to groups of neurons. That is the prime consequence of plasticity: sensitivity to input.

From the perspective of brain neuroscience, the environmental world of the developing embryo and fetus is as important as any information carried by genes. If you like machine analogies, we are self-wiring machines—at least the most important part of us is: the cerebral cortices of our brains. We continually build and rebuild the connections in our brains, and the self-wiring starts when the first cortical nerve cells migrate, differentiate, and start connecting to each other. Of course, much of the early self-wiring is evolved and heavily dependent on evolved genes and evolved gene expression and maybe a degree of randomness. But once the general plan of the cerebral cortex is laid down, it's the input to the cortex from inside and outside the body that essentially shapes the connections within the cortex.

Connection plasticity, as mentioned, makes learning, memory, and responsive behavior possible. But its most important characteristic in the context of this book is its nature: Both the quality and quantity of cerebral cortex plasticity are vulnerable to environmental effects that occur during development. Aside from altering plasticity, these effects can also alter developing connections themselves, producing changes in behavior ranging from that associated with subtle personality traits to that associated with psychopathology.

For obvious reasons, there's great clinical interest in identifying genes that may "underlie" various human behavior disorders. But this effort is problematic from the get-go because, from a genetic standpoint, behavior and behavior disorders are non-Mendelian complex traits, meaning that they likely arise from gene complexes rather than from single genes.[6] In addition, the term "underlie" is ambiguous. If we remove the tires from an automobile and examine the consequence, is it useful to promote the idea that the automobile tires "underlie" the automobile behavioral trait called "steerability"—the trait that allows the automobile to be steered in one direction or another? Relatively few genes have been identified that are directly responsible for complex behaviors of any kind, disordered or ordinary, and the reasons for this are clear to professional geneticists: locus heterogeneity, low penetrance, variable gene expression, pleiotropy, and epistasis. Locus heterogeneity refers to the variability of gene location in chromosomes; penetrance refers to the degree of expression of a dominant trait, which can vary from one individual to another; pleiotropy refers to an involvement of a gene with more than one trait; and epistasis refers to interactions between genes, for example, one gene affecting the expression of another gene.

Important brain connections are formed during development, both before and after birth. Primates, including humans, have relatively long gestation periods, and we can speculate about the consequences of this for development of the brain.[7] Although the question of how much of the details of brain development are a product of evolution and how much a product of environmental interactions is far from a settled issue, it's clear that any idea of a filled slate at birth is no more acceptable than the idea of a blank slate at birth. Rather, according to evidence from modern neuroscience, the elaborate architecture of the adult human brain is the final product of gene-based interactions, global and specific cellular interactions, and ongoing interactions between the developing individual and the environment. In plain language, everything is involved in the development of brain tissue, and at the present time it makes no sense to say that one origin of that developmental process is more "important" than another origin.[8]

BRAIN-DEVELOPMENT PROCESSES

The prenatal development of the nervous system below the cerebral cortex, like the fetal development of other organ systems, involves a dynamically organized program. The idea of "organizers" during embryonic and fetal development was first introduced in nineteenth century embryology. In general, an embryonic organizer is a part of an embryo that induces another

part to develop. The concept derived from experimental observations of sequential development, especially in animals from
which a specific embryonic tissue had been removed. Some tissues were found to be essential for the later development of other
tissues—even if in the nineteenth century hardly anything was
known about the sources of influence. These days, organizers
are said to be parts of the embryo that induce cellular differentiation of other parts—which means changes in gene expression.
These concepts, coupled with modern molecular biology, have
been found useful in understanding the development of various
parts of the brain.[9]

Modern developmental neuroscience presents us with a story
of how the brain develops; a story whose details have been accumulating during the past 50 years. What we need to know concerns organization. If the organization of the brain determines
its function, the central developmental question is how does the
brain get organized? In the context of this book, that question
is paramount, since any environmental impact that affects organization will affect behavior and cognitive performance. Fifty
years ago, a discussion of brain organization during development could focus only on large brain structures because next to
nothing was known about how nerve cells get organized, how
they move and how and why they connect with each other. Now
we know some details, and here are some that show us possible important targets for environmental impacts on fetal brain
development:

- Evolution by natural selection requires variation, differences between cells, organ function, individuals, or groups. Many important phenotypic variations of the human brain are developmental variations. The variation required by natural selection operating during prenatal development occurs primarily as a result of random and nonrandom local environmental effects on gene expression. The importance of this for our context is that it tells us the idea that fetal development is malleable by circumstance is expected from what we already know about the evolution of development.

- The neurons in the brain are not all of the same type. They differ in their shape, their biochemistry, how they receive input from other neurons, how and to where they transmit output. The various final types of nerve cells in the fetal brain have lineages, origins in prototypes. Developing lineages proliferate, migrate, survive, and differentiate in patterns specific in time and space. The patterns depend on cellular signals. The development of any lineage also depends on the development of other lineages. Some lineages may be more vulnerable to damage than other lineages. Damage to one system can cascade through many systems.[10] The relevance for our context is that depending on the nature of specific impacts on the fetal environment, effects can be specific, impacts affecting only one or a few cell lineages. It's not necessary that a particular impact on the fetal environment always have a global effect on the entire brain.

- During the past decade, hundreds of studies have demonstrated the importance of epigenetics in the developing brain—the interaction of local environments with gene expression. Most of these studies have focused on the critical periods during which postnatal brain development in infants and children is most vulnerable to environmental effects.[11] Although similar studies of human fetal brain development are relatively crude, they do show that the environmental effects on the developing fetal brain differ in their consequences depending on how early or late they occur. The consequences are the windows of vulnerability we have mentioned previously. Such windows make biological sense—their absence would be remarkable in any developing tissue.

- The two chief features of fetal brain development are the migration of nerve cells to particular places and the formation of neural circuits after they arrive there. Although once neurons arrive in a location, their place is apparently more or less fixed, the connections they make, the circuitry, the wiring, is not fixed at all. There's plenty of evidence that demonstrates that neural circuits in the cerebral cortex are dynamic, malleable, changing with activity, connections between nerve cells made and unmade. During development of the brain, connections are as malleable to impacts as migrations of nerve cells. Since connections and cell density in various regions of the brain are most certainly related to later behavior and cognitive performance, we should not be surprised when we find later behavior and IQ correlated with impacts on the fetal environment.

- In all of the mentioned factors, the cellular migrations and organization of the cerebral cortex during fetal brain development are critical. But evidence for the neuronal migration that precedes the formation of neural circuitry in the prenatal human brain is not easy to come by because there are no ethical techniques for examining this process in humans. However, techniques do exist for animals. Real-time imaging of animal (mouse) cortex slices reveals that in brain development early and late migrating neurons use different types of migration mechanisms.[12] Two distinct modes of cell movement exist. During the early stages of the formation of the cerebral cortex, neurons move without guidance by translocation of their cell bodies. Later, migration of neurons involves both free translocation and movement along the extensions of networks of supporting cells (glial cells). No matter how cells move, such movements are usually in response to chemical gradients, and that means any local environmental change that alters the local chemical gradients outside migrating cells can have an impact on the organization of the fetal brain.

- What happens inside a nerve cell when it moves? The intracellular mechanisms of nerve cell movements are complicated. First an extension of the neuron is sent out, and then the neuron nucleus follows the extension.[13] This sort of process depends on a network of special macromolecules (microtubules) in the neuron cytoplasm that form a cage around the nucleus and essentially drag it forward. Any chemical impact that affects this microtubule structure will seriously affect neuron migration. So here's another point of vulnerability in the nerve cell migrations that position nerve cells at functionally specific locations in the developing brain.

• It helps to think of the human brain as divided into an evolutionary "old brain" and an evolutionary "new brain"— the part of the brain evolved more recently in higher mammals. The new brain is the cerebral cortex; everything else is old brain. Although hard-wiring does occur in the evolutionary old spinal cord, cranial nerves, and autonomic nervous system, and also in the evolutionary old brain of the developing child, there's very little "hard" wiring in the evolutionary new brain, the cerebral cortex, especially in those parts of the cortex that are associated with learning and memory. This means the developing new brain is a malleable tissue, sensitive to its inputs, sensitive to its local environment, and ultimately sensitive to the entire uterine environment.

• This anatomical difference in plasticity between the evolutionary old brain and new brain is of great importance. The so-called higher functions in behavior—perception, analysis, thinking, language acquisition and use—are all primarily products of activity in the new brain—the cerebral cortex. In contrast, "old-brain" behaviors—rage, fear, sexuality, responses to hunger and thirst—are for the most part hard-wired during development of the brain before birth, with some continuing development during childhood. But this does not mean that primitive behaviors are not malleable during fetal development. The cerebral cortex ordinarily rules voluntary behavior, exerts executive control over the old brain. Sometimes the control is not enough and sometimes it's more than necessary. Whatever affects new-brain control can affect old-brain function—and such effects can result from impacts during fetal development.

Such are some of the details of what is happening during the development of the fetal brain. Although only a few millimeters in thickness, the adult mammalian cerebral cortex is highly organized into layers that differ in neuron density, neuron type, and the architecture of connections. The adult human cerebral cortex has six layers (birds and reptiles have only three layers), and during development neurons must migrate to their proper place and establish the connections that determine critical cortical functions. These connections are made in response to local chemical signals and inputs from other neurons near and far. Any defects in this process can cause various developmental disorders of higher brain function.[14]

A "developmental disorder" is a recognized clinical consequence of an aberration during prenatal development. But what of unrecognized developmental aberrations, variations too subtle to be clinically significant but significant enough to affect later behavior and intelligence? Do such "subclinical" variations exist, and if so, when and how do they happen?

Unfortunately, for the development of the human brain, this question can be addressed only indirectly—from animal experiments and by deduction. But even animal experiments are problematic because it's not always clear how measurements of animal behavior have any serious meaning for measurements of human behavior. An animal, even a chimpanzee, is not just a simplified human being exhibiting simplified human behavior. Deduction, on the other hand, gives us a clearer answer: It stands to reason that clinical observations of the effects of brain-development aberrations are the proverbial tip of the iceberg—with an entire universe of subclinical differences, variations, and developmental aberrations that we have no means yet to measure.

The important point here is that the production and migration of nerve cells are provoked by local biochemical events easily affected by the local prenatal environment, and the same is true for axon guidance processes that connect nerve cells to other nerve cells—connections of critical importance for later cognitive and emotional behavior.

A point about the life and death of nerve cell connections: in recent years, it has become apparent that during brain development the formation of many neural circuits involves selective elimination of many synapses—selective pruning. This is an old idea reinvigorated in a theory of "neural Darwinism,"[15] which in general proposes that groups of synapses are selected for survival or extinction according to experience—a model for learning that has some observational evidence to support it. Significant pruning of synapses evidently occurs during childhood, and it's apparently associated with learning behavior. Moreover, according to an important new finding, proteins that function as part of the immune system may also be involved in pruning synapses, with certain synapses tagged for elimination by an immune system protein called complement. This suggests that immune system proteins may be implicated in certain neurodegenerative diseases that destroy synapses.[16]

It's estimated that each cortical neuron has an average of 10,000 synapses with other neurons near and far. But these connections do not develop all at once. They are formed in the brain during fetal development, infancy, childhood, and even adulthood. The simple fact that the gross anatomy of a child's nervous system is not completely developed until about 11 or 12 years of age is another refutation of the idea that neurological wiring is complete at birth and therefore uninfluenced by environment.

So the first principle is that neural development does not stop at birth, and the second principle is that from the beginning of development in the embryo to the very end of adult life many connections between nerve cells in a particular region of the cortex and between that region and other regions of the brain are responsive to inputs of neural activity and are constantly changing. In other words, the human brain is not a static analytical system but a dynamic analytical system—one whose microorganization depends on what happens to it. Psychologists use the term "learning" to refer to the acquisition of behavior patterns based on experience. The human brain, in effect, is "learning" throughout its life history.

We know that in the mature brain, neurotransmitters are released by nerve cell endings and act at synapses to mediate the interaction between nerve cells. There's now evidence that during the development of the prenatal brain, neurons can release transmitters before any connections are made, and that such released transmitters, acting as "trophic factors," guide the formation of connections.[17] Any environmental impact or mutation that affects the synthesis or release of neurotransmitters can be expected to have an effect on the development of the fetal brain. For example, during fetal development in Down syndrome, reductions apparently occur in the levels of various neurotransmitters. This may be one mechanism for the impaired brain development characteristic of this syndrome.[18]

Another developmental disorder is called lissencephaly, the term for disorders of brain formation in which the surface of the cerebral cortex appears smooth and the layers of neurons in the cortex are abnormal.[19] The brains of patients with lissencephaly have only four layers rather than six. As you might expect, these patients are profoundly retarded from birth, often develop seizure disorders, and in severe cases, have abnormalities in the heart, kidneys, and other organ systems. In 1993 researchers discovered that these patients are missing one copy of a gene in chromosome 17, the gene now called LIS1. The cause of the missing gene is evidently a microdeletion, a mutation that results in the loss of a stretch of DNA in a chromosome. The larger the deletion, the more severe the consequent disorder, which probably means that nearby genes are affected by the larger deletions. What is known about this gene is that it does not act alone, and that it's involved in the regulation of nerve cell division, nerve cell maturation, and nerve cell migration.[20] Unfortunately, there's no animal model yet that might tell us more: mice and rats already have a smooth brain surface and are not useful as experimental animals for this developmental disorder.

What does all of this tell us about the vulnerabilities of brain development? Here we have a specific gene whose absence causes serious aberrations in the structure of the developing cerebral cortex, which aberrations in turn cause profound mental retardation. It's obvious that the gene must be expressed at least sometime during brain development—or its absence would be irrelevant. A microdeletion eliminates the gene and causes pathological brain structure.

But what happens when rather than elimination of this gene, the regulation of its expression is changed? Are subtle changes in cortical structure a possible result? Can the expression of this gene be affected by changes in the fetal environment? My point is that the fact that specific gene mutations can be associated with specific clinical disorders is no reason to ignore the idea that the fetal environment can seriously affect gene expression during brain development. For every gene mutation identified as related to a severe brain disorder, we need to ask what happens when it's the expression of that gene that's changed rather than its DNA code. The answers are critical for our understanding of how subclinical behavior and behavior differences between individuals can be epigenetically affected.

What all these technical details mean for us can be briefly stated in plain English: We know enough about the detailed development of the human brain to expect it to be highly vulnerable to impacts on the fetal environment. Neuroscientists are not surprised by correlations between impacts on the fetal environment and later behavior and cognitive performance. The purpose of this chapter has been to explain this lack of surprise.

FIVE

LIFE IN UTERO: SHAPING OR DESTRUCTION?

During the Second World War, the Netherlands was occupied by Germany, and in the fall of 1944, in retaliation for a Dutch railroad strike to aid the Allies, the Nazis cut off food supplies to the population on the western part of the Netherlands.[1] For the next 9 months, a major fraction of the Dutch population lived on less than 1000 calories a day, eating anything at hand, including tulip bulbs, to survive. The people involved suffered from chronic hunger and diseases produced by malnutrition, and ultimately 18,000 Dutch people starved to death. Comparing populations, this would be the equivalent of more than 600,000 people dying of starvation in America.

Later, this tragedy, known as the Hunger Winter, came to have a unique scientific significance for several reasons[2]: (1) the famine was sharply circumscribed in time and place; (2) the affected population had major problems obtaining food elsewhere, so the famine conditions were relatively constant; (3) the population was ethnically homogeneous and without marked prior differences in dietary patterns; (4) the official food rations were known for weekly periods, so that the number of calories available could be estimated by place and time of birth; (5) the availability of food was largely unaffected by social class; and (6) long-term follow-up was possible, since individuals in Holland could be traced through national population registers.

The Dutch Hunger Winter thus provided science and clinical medicine with a population of pregnant mothers who had experienced malnutrition in first, second, or third trimesters. Researchers also had access to the later childhood and adult medical histories of the fetuses that survived the famine, and similar histories of the children subsequently born to those fetuses. No famine has ever had its transgenerational consequences so carefully tabulated and examined. Among the findings was an increase in the prevalence of schizophrenia and other mental illness among the offspring in gestation during the famine.

The Dutch Hunger Winter, in short, provided one of the major sources of evidence for a new idea in the understanding of later health and disease. This idea forms the hypothetical background for this chapter's focus on specific factors inside and outside the fetal environment that influence, and sometimes disrupt, fetal development.

The idea is "fetal programming." Fetal programming and its variants—such as the Barker hypothesis, the fetal or developmental origins hypothesis, the developmental origins of disease hypotheses, and some others—all refer to the same phenomenon: An interaction of the environment with the fetus whose effects may not show up until years later and may or may not be transgenerational. Apart from this common thread, however, there are some significant differences in the emphasis of each hypothesis.

For example, epidemiological findings with humans and experimental studies with animals have demonstrated that individual tissues and whole organ systems can be "programmed" (scheduled) during critical periods of fetal development with significant consequences for the function of these tissues and organ systems in later life. In the 1980s, British epidemiologist David Barker shocked the biomedical community by presenting massive evidence that low birth weight was correlated with coronary artery disease, hypertension, stroke, and type 2 diabetes in adults.[3] He hypothesized that prenatal alterations in fetal nutrition and endocrine status result in developmental adaptations that permanently change structure, physiology, and metabolism, thereby predisposing ("programming") individuals to later cardiovascular, metabolic, and endocrine disease in adult life.[4] Researchers have proposed extending the Barker hypothesis to include the impact of prenatal toxic chemicals on brain development and the subsequent appearance of neurodegenerative diseases, such as Alzheimer's disease and Parkinson's disease.[5] The reasoning is simple: The brain is just as biological as any other part of the human body, and it is therefore subject to impacts on fetal development and fetal physiology.

The term "fetal programming" is thus also used to signify a much broader range of "programmed causes" and their later effects—that is, to encompass not only predispositions of individuals to later physical disease, but also predispositions to later behavior patterns that range from ordinary to pathological.

In this book I use the term "fetal programming"—without reference to mechanism—to describe environmental impact of any kind on the fetus that produces alterations in health and disease apparent at any time in childhood or adulthood. This idea is essentially an extension into the prenatal period of more general conclusions about the role of early experience in later life. Included in this conception are postnatal temperament, behavior patterns, intelligence, psychopathology, and so on—the behavioral focus of this book. The point is that sometimes postnatal consequences of prenatal environmental impacts involve more than just alterations in the physiology of organ systems—such impacts can also have consequences for later behavior and cognitive performance.

During fetal development, there are critical periods of vulnerability to environmental impacts, the critical periods occurring at various times in different tissues. Cells rapidly dividing to form tissues and organs are at greatest risk, and it's reasonable that fetal programming is more likely to occur in such tissues.

But human research in this area nearly always involves correlations rather than controlled experiments that relate effects to causes, and as a result inferences about human prenatal programming usually have serious limitations.

Moreover, not much in development is strictly predetermined or permanently fixed. Epigenetics has random elements and so is essentially probabilistic, and gene expression itself is activity-dependent. Nowhere is this developmental dynamism more evident than in the cerebral cortex of the human brain. It's inconceivable that the entire cortex is functionally wired up by genes without significant dependence on neurological input during and after gestation. Such a tissue would have no plasticity and no evolutionary advantage. Without a plastic cerebral cortex, we would be no more than ants—large ants, but hardly better than automatons. (As you might expect, ants do not have a cerebral cortex or even anything functionally equivalent to a brain. As far as we know, there's hardly any plasticity in the ant nervous system—which is why the behavior of ants today is not much different than the behavior of ants a million years ago.)

The biologic mechanisms underlying fetal programming are poorly understood. It seems a reasonable idea, but researchers need to forge ahead and gather more evidence to support it. Here, too, correlations are suggestive, not a demonstration of cause and effect. Maybe the first contribution of the fetal programming idea is the new attention to the possibility of nondestructive fetal impacts as variables in explaining various postnatal behaviors.

INSIDE THE FETAL ENVIRONMENT— VULNERABILITY REVISITED

As I pointed out in chapter 3, one of the ways to regulate gene expression is to switch off genes by DNA methylation or switch them on by demethylation. DNA methylation patterns in ordinary (somatic) differentiated cells are usually stable, their methylation patterns are passed on to daughter cells when they divide.

But during early gestation, mammals have at least two developmental periods—in germ cells and in preimplantation embryos—when a substantial part of the genome is demethylated to begin cell differentiation from scratch.[6] One of the critical features of embryonic development is the reprogramming of DNA methylation patterns essential for cell differentiation into various tissues, and of course any environmental impact that affects this process will affect tissue development—including the development of various cell clusters in the brain. Certain toxins, such as polyphenolics (antioxidants derived from plants), are already known to affect DNA methylation by chemical modification of its target, the DNA nucleotide base cytosine. As we learn more about the detailed chemistry of DNA methylation and demethylation, we will know more about other environmental toxins that have the potential to make subtle changes in that chemistry.[7] Indeed, I suspect that an enhanced understanding of the mechanism of DNA methylation will turn out to have a profound importance in both evolution and development. If DNA methylation patterns are responsible for cell differentiation, where is the information about patterns stored, and what triggers the institution of a methylation pattern in the DNA of a particular progenitor cell? Unraveling this mystery may ultimately be as important as the past unraveling of the genetic code for proteins.[8]

In an ordinary pregnancy, after fertilization in one of the fallopian tubes, the fertilized ovum (zygote), which now contains a full complement of chromosomes, divides into two cells in about 30 hours, into four cells in about 40 hours, and into 16 cells in about 3 days. While this is happening, the new preimplantation embryo is traveling down the fallopian tube to the uterus. At about 3–4 days, in the late morula stage, it will enter the uterus. During this period, important epigenetic modifications of the embryo's genome are occurring at a rapid pace.

Despite its autonomy, the preimplantation embryo, from the time of fertilization of the ovum to the time of entry into the uterus, is highly influenced by its external environment. Many preimplantation embryos already have their development program corrupted by genetics or environment or both and are not viable when they enter the uterus. Other embryos, for a variety of reasons, never achieve implantation in the uterine wall and are soon destroyed. The exact percentage of seriously defective early embryos is unknown. These embryos are usually lost before a woman realizes she's pregnant and the lost embryos remain undetected. The current estimate is that 50 percent of pregnancies end in spontaneous abortion. Only about half of these losses are the result of chromosomal abnormalities.[9]

There is no evidence to argue that the other half result from prenatal environmental effects. But it's reasonable to assume that the prenatal environment in its various forms can be an important factor, and that countless possibilities exist for both lethal and nonlethal effects due to that factor. The simple fact that so many embryos are apparently lost means embryos are particularly vulnerable to damage.

The most important general idea about the fetal environment is that any impact that results in physiological stress for the fetus can be a risk factor for adverse developmental and health outcomes. This is clear only from animal experiments. For humans, on whom comparable experiments are not possible, we must depend on empirical data, on correlations, and on deductions. Moreover, because the physiology, chemistry, hormone profile, and so on of the embryo and fetus are continually changing during gestation, the developmental consequences of an environmental impact depend on when it occurs, and on its duration, severity, and type.

The placenta, which forms after implantation, is a complex tissue system that acts as an intermediary between the mother and the fetus. It apparently functions as a nutrient sensor, directly regulating fetal nutrient supply and fetal growth and thereby playing a central role in fetal programming. In addition, perturbations in the maternal environment can affect the methylation status of placental genes and result in changes in placental function. When the placenta is dysfunctional enough to reduce transport of nutrients, the result can be fetal growth restriction, deficits in neural connections, and deficits in the myelination of nerve fibers.[10]

The role of the placenta is of critical importance in the development of twins. With identical (monozygotic) twins, even a shared placenta does not guarantee uniformity and equality of the fetal environment.[11] Each twin sharing a placenta is supplied by a different portion of the placenta. The sharing is almost always unequal. Delivery of maternal blood differs. The amount of space and the location of each fetus differ. Each twin has a local environment in addition to a general uterine environment.

Epigenetic Inheritance

As in every other biological system, fetal mechanisms to sustain survival by compensating for various environmental effects do exist. But as with other biological systems, survival mechanisms often involve a trade-off between a severe outcome and a lesser outcome. The fetus survives, but with a mild consequence recognized only as "phenotypic variation"—a difference between individuals that results not from genetic differences or postnatal environmental differences, but from differences in prenatal environments.

One of the essential features of epigenetics is that it provides a nonmutation environmental explanation for this phenotypic variation. Since the differentiation of tissues and organs of multicellular biological systems involves selective and relatively stable regulation of the expression of various genes, and given that gene expression and regulation are ultimately controlled by environmental influences, many differences in phenotypes can be explained as the result of different experience differentially activating or changing levels of gene expression.

From all of this comes the concept of "epigenetic inheritance"—an idea that in its broadest sense encompasses all phenomena in which the transgenerational transmission of phenotypic variation occurs without variation in DNA base sequence—inherited changes without gene changes. Epigenetic inheritance across generations is relatively common in plants, but it's still unclear how widespread this phenomenon is in mammals, or how much of a role it has had in shaping evolution.

Epigenetic mechanisms in general may explain paradoxical findings in human twin and inbred animal studies when phenotypic differences occur in the absence of apparent postnatal environmental differences. That's the key idea: Identical genomes do not imply identical gene expression, which means phenotypes can differ even with the same genotype. In principle, the corollary is also possible: Similarity in phenotypes can be produced by appropriate gene expression in different genotypes. It's not genes that rule, its gene expression that rules—and gene expression can be shaped by environment.

Nor are the brain and its output exempt from these mechanisms. In the developing brain, as suggested in the previous chapter, neuronal activity-dependent regulation of gene expression may specify neuronal fate and function.[12] We may not yet have a superabundance of evidence to support this epigenetic idea, but human genetic studies have linked several transcription regulators to neurodevelopment disorders, including mental retardation. We can assume that subclinical epigenetic effects also exist that modify gene expression required for synaptic development during development—and this should affect later behavior and cognitive performance.

The classical (and simplistic) idea in genetics has been that the genotype (the total repertoire of genes inherited by the individual) produces the phenotype (the total repertoire of observable traits) of the individual. But with the advent of modern biochemistry, physiology, molecular biology, and so on, it has become more useful to speak of the "exophenotype," the repertoire of traits that are observable, and the "endophenotype," the repertoire of traits that are not (yet) observable. For example, the growth of horns in a male moose (exophenotype) is observable, while some maybe yet unknown hormonal complex (endophenotype) responsible for the growth of male moose horns is not observable—although both may derive from the genotype. A more narrow definition of the two types is clinical and often psychiatric: An apparently inherited behavioral dysfunction or psychiatric illness is an exophenotype, while an as yet unknown, inherited, and nonobservable neurotransmitter defect is an endophenotype.[13] Note that this psychiatric definition denotes an assumption of genetically-determined intermediates between genes and behavior, but so far largely neglects the concepts of epigenetics and the idea of gene–environment interactions.

In any case, the early fetal cerebral cortex is not yet specialized in function, which allows its interaction with the fetal environment to play a crucial role in gene expression and in the ultimate cognitive phenotype. Indeed, during the past decade, evidence has accumulated for the idea that prenatal environment impacts can lead to many of the neurological and behavioral deficits that become apparent only after birth.[14] The timing and severity of these impacts are critical. Clinicians, of course, focus on clinical effects. But we can assume that depending on timing and severity of an impact, effects on the developing brain can be subtle enough to be subclinical but still produce phenotype behavior variations. Subtle brain injury can occur to a particular class of neurons with significant effects on the function of a specific brain system.[15]

The classical approach to the development of behavior involves four sets of variables: genetics, fetal developmental variables, parental (or family) impact on the child, and cultural impact on the child. The science of epigenetics now suggests another set of variables—epigenetic inheritance, the transmission to offspring of parental phenotypic responses to environmental impact—even when the young do not directly experience the impact themselves. Pathways for such interactions have been identified. For example, maternal stress may affect not only a daughter's gestation but also the gestation of the daughter's offspring. Temperament variation may be influenced in the same way—transgenerational non-DNA inheritance of temperament.[16]

One other variable is worth mentioning: the season of an infant's birth. We will deal with this again in a later chapter. High levels of sunshine, as in summer, may increase the fetal level of insulin-like-growth factor (IGF)-1, a polypeptide protein hormone similar in molecular structure to insulin, and facilitate prenatal growth. In any case, exposure to peak sunshine during the first trimester correlates with heavier birth weight.[17] The most probable mechanism is a change in placental metabolism during the period when placental growth is most rapid.

In addition, anthropologists have analyzed the long-term effects of early developmental conditions in a population in which both nutritional levels and the burden of infectious diseases show a seasonal variation. For example, a study of a village on Minorca Island, Spain,[18] matched birth and death data for 4646 individuals born between 1624 and 1870. The results show that the season of birth had a long-term effect on survival in the birth cohort 1800–1870: summer births had a lower risk of death after age 15. One explanation for the results is a lower susceptibility to degenerative diseases in adult years due to superior in utero nutrition for summer births—a fetal programming hypothesis.

On the other hand, as discussed in later chapters, there's also considerable evidence that in industrialized countries an individual's season of birth may influence susceptibility to a wide range of psychological disorders, including schizophrenia, eating disorders, and substance abuse.

MATERNAL–FETAL COMPETITION

We tend to imagine that during gestation the mother and fetus function as a cooperative unit, but as I've alluded previously, that's not always the case. The body and nervous system of one often works against the body and nervous system of the other. The changes in maternal physiology during pregnancy, for example, involve multiple adaptations of the mother both to protect her own health and to optimize fetal growth and development. Many of these adaptations are organized by the maternal brain and mediated by specific hormones of the neuroendocrine system.[19] But some physiological adaptations (for example, maternal regulation of maternal calcium and iron metabolism) necessary for the mother's health during pregnancy may compete with the developmental needs of the fetus. Perfect harmony is not always the rule (and disharmony is sometimes dangerous for the mother, for example, preeclampsia). The outcomes of these competitions are of tremendous importance for the later behavior of children and adults.

Other factors within the maternal environment may also result in conflicts with and significant disruptions of fetal development, among them:

- *Placental insufficiency*. From a biological standpoint, the placenta is a filtering and hormone-secreting organ that mediates the interaction between mother and fetus. A damaged placenta causes a damaged fetus. Placental insufficiency, for example, is a pathological condition consisting of a placental functional deficit; its cause or causes are usually unknown.[20] Among its possible consequences are morphological changes in the fetus, and it's apparently the primary cause of what clinicians call "intrauterine growth restriction" (IUGR) or "small for gestational age" (SGA), terms that describe a fetus whose length and weight do not correspond with the calculated age of the fetus in months or weeks. The diagnosis is usually made if there's a combination of fetal growth restriction, decreased amniotic fluid volume, and abnormal ultrasound measurements. The important parameter is gestational age: An infant who is "small" because of a premature (preterm) birth may not be a case of IUGR. The 10 percent of all babies that do show some degree of IUGR are statistically at increased risk of neurological and other problems.

- *Nutrition.* Fatty acids in the maternal diet are critical chemicals for the embryo and fetus during growth and development. These molecules are precursors to cell-membrane fat molecules (lipids) and they are also specific binding and interacting molecules (ligands) for receptors and transcription factors that regulate gene expression. Inadequate fatty acid nutrition during fetal growth affects the chemistry of nerve cell membranes and can disrupt the migration of neurons in the fetal brain.[21] Any chemical impact in the environment that affects fatty acid chemistry can have significant effects on neuronal circuitry.

- *Hypoxia.* Oxygen supply is as important as nutrient supply in fetal growth and development. The association between low birth weight and clinical problems in the adult phenotype has been linked to poor nutrition and/or poor oxygen supply (hypoxia) during gestation. It is also well known that hypoxia induces fetal malformations in a great variety of experimental animals. Fetal hypoxia may be transiently produced by temporary occlusion of the umbilical cord, and by chronic deficits in oxygen due to maternal hypertension, tobacco smoking, and partial placental detachment. Even relatively brief periods of fetal hypoxia can result in neuronal death (in cerebellum, hippocampus, and cerebral cortex), white matter damage, and reduced growth of neural axons and dendrites.[22] These effects are more profound at mid than at late gestation.

- *Temperature changes.* Another common gestational problem is high maternal body temperature that translates into high fetal temperature. Fetal hyperthermia can result in severe consequences: anencephaly, spina bifida, mental retardation, facial defects, cardiac abnormalities, and limb defects—all classical clinical abnormalities.[23]

- *Maternal infection.* Mild infection transmitted to the fetus is another source of possible subclinical changes in brain development.[24] Clinical and epidemiological studies have shown associations between maternal infection, fetal brain damage, and increased levels of proinflammatory cytokines, a class of regulatory proteins secreted especially by cells of the immune system, in amniotic fluid and the fetal brain. If microorganisms gain direct access to the amniotic cavity and the fetus, they induce the innate immune response and cause inflammation of the chorioamniotic membranes and the production of proinflammatory cytokines that soon pass into the fetus by various routes.

- *Inflammation.* Exposure of the fetus to inflammatory agents can cause brain damage, especially in the white matter of the nervous system, and the effect is increased by hypoxia.[25] Inflammation usually follows infection.

- *Glucocorticoids.* Exposure of the fetus to the class of steroid hormones called glucocorticoids (for example, cortisol) during prenatal development has been proposed as one of the principle programming factors for increased risk of chronic diseases among individuals born with low birth weight—and it's associated with an increased probability of later development of hypertension, diabetes, and psychiatric disorders, such as depression and anxiety. I will discuss some of these connections in greater depth in Chapter 11.

Still other competitive factors associated with maternal behavior or exposure to toxins can cause similar disruptions to fetal development as well, such as

- *Tobacco use.* Cigarette smoke inhaled by a pregnant woman affects the fetus through the direct effects of nicotine and other chemicals, hypoxia associated with increased blood levels of carboxyhemoglobin, nicotine-induced constriction of placental blood vessels, and a general decrease in the flow of nutrients and oxygen from mother to fetus. Long-term consequences for offspring behavior, according to magnetic resonance imaging (MRI) analysis, include thinning of the cerebral cortex, especially in female adolescents, and lower birth weight in both sexes of subjects whose mothers smoked during pregnancy.[26] Second-hand smoke inhaled by a pregnant mother should in principle cause similar damaging effects to her fetus.

- *Cocaine use.* Like alcohol and other pleasure-drugs, cocaine use is an American endemic plague. Among pregnant women, rates of cocaine use vary from 1 or 2 percent in rural localities to as much as 18 percent in the urban inner city of places like Boston.[27] In the fetuses of these women, prenatal cocaine exposure may alter brain development through changes in neurotransmitter systems[28] and is associated with postnatal developmental deficits, including increased physiological arousal and altered physiological responses to sensory challenges as measured by heart rate, heart rate variability, and cortisol levels.[29]

- *Pharmaceuticals and other substances.* There are categories of pharmaceuticals and other substances whose damaging effects on embryonic and fetal development are well known, suggesting that the developing embryo and fetus are highly vulnerable to unusual chemical change. These agents or substances include synthetic retinoids, sex hormones, anticonvulsants, vaccines, thyroid drugs, narcotics and analgesics, various other psychoactive drugs, antibacterials, anticoagulants, cardiovascular drugs, and excessive caffeine (more than seven or eight cups of coffee per day).

Evolution may have prepared the development program to deal with toxin molecules "known" to the evolutionary process, but there's no preparation to protect the developing embryo or fetus from new toxin molecules, such as tobacco, cocaine, or other substances that were never part of an ancient cellular environment. Ethanol, the alcohol in alcoholic beverages, is a special case. An ancient molecule found everywhere in ripe fruit and certain other natural foods; it is nonetheless toxic to the embryo and fetus. It's a small molecule whose chemistry makes it easily soluble in both oil and water. As a consequence, it's almost impossible to protect cells against ethanol once they are directly exposed to it. Ethanol passes rapidly between and through cells, and it can interact in many different ways with large and small biological molecules active in a biochemical cascade. Even low concentrations of ethanol, which may be safe to an adult, are highly poisonous to an embryo or fetus.

THE CHALLENGE OF PRENATAL
TOXICOLOGY

The phrase "low concentration" can be deceptive. A toxin concentration too low to be measured by most clinical instruments can still involve 100 million toxin molecules per cubic

centimeter of blood. We mustn't think that a low or even undetectable toxin concentration means only a few toxin molecules
are around to do damage. What governs the toxicity of a toxin
molecule is really not its concentration but the nature of its
chemical interaction with a biological system. If a toxin molecule interacts with DNA, only a few hundred molecules interacting very early in gestation may be enough to cause a gene
mutation ultimately lethal for the embryo or fetus. Other interaction variables are the rate at which toxin molecules are broken
down by biochemical reactions into harmless components, and
the rates at which toxin molecules are adsorbed or sequestered
into cellular compartments and so prevented from interacting
with critical cellular structures and processes.

Researchers naturally search for useful generalizations about
toxins and their toxicities, but toxins of various types have different chemical structures and chemical reactivities, and developing biological systems are not the same as adult biological
systems. So generalizations sometimes require caution. For
example, a central and old concept in toxicology is that every
toxic substance has a dose below which there is no detectable
toxic effect—an outcome of the so-called dose–response relationship paradigm. But this concept may not be useful for toxins
that affect developmental cascades—and we have two very good
examples of this in lead and ethanol. Many researchers now view
fetal toxicity as a domain of a different sort—a domain where
classical ideas about toxins and toxicity may not be applicable.
Every case in which a pregnant woman is unharmed by a toxin

that damages her fetus is a demonstration of this new principle. The other important caveat about toxicity is that unless you know precisely what effect to measure, there's no way to even think about a dose–response relationship.

Molecules live in a special world. Given a chemical compound that has a blood concentration of 10 micrograms per deciliter (the current poisoning "threshold" concentration for lead), a single drop of that blood may contain 10 trillion molecules of that chemical. If the chemical is a toxin, how many molecules does it take to cause damage? We don't know. We can test the toxin on animals and determine a concentration below which we don't see some effect. We can test the same toxin on cells in culture in the same way. One problem is what "effect" do we look for? What "damage" do we look for? Sometimes we need to know precisely what the toxin does to know how to definitively measure any effect or detect when an effect is no longer present. Unfortunately, for most neurotoxins, we don't know the molecular details that make them toxic. What we can see as "damage" is limited by our instruments and by our understanding of the biological system. Tomorrow, damage that we can't see today may be visible with new instruments. In the nineteenth century, early lung damage by coal dust was essentially invisible. In the twentieth century, medical chest x-rays and computer tomography made such damage routinely obvious.

Now suppose the toxin target is a single human ovum or an embryonic human morula of sixteen cells. Can we determine how many molecules of that toxin are necessary to derail that ovum or morula to an aberrant developmental trajectory? If we're allowed to manipulate the environment of the embryo, we could in principle make serial dilutions to produce extremely small concentrations of a toxin and do experiments—introduce the small concentrations into the environment of the fetus. Without such manipulations, all we can do is try to measure concentrations derived from the environment—which means our determination of concentration sooner or later vaporizes because our measuring instruments are limited in their resolution. We have no instruments that can detect a thousand molecules of a specific compound in solution, let alone a few dozen. We cannot determine how many molecules are necessary to damage a morula, and we cannot detect everything happening in the morula anyway.

So we resort to logic. Is it possible that only a few molecules of a toxin entering a cell can start an amplifying chemical chain reaction that produces a large effect on that cell? It's certainly possible in principle. We know that a single α particle produced by radiation can cause a mutation in DNA, but we have no idea about the possible damage by a single molecule or of a few molecules or of a few thousand molecules of a specific toxin that gets to an ovum or blastocyst or morula or embryo or fetus. Again, defining "damage" is a critical problem. Without knowing what effect to measure, any definitive assessment of toxicity is not possible.

Welcome to the bizarre and foggy world of prenatal molecular toxicology. There's really no hard-boiled rational basis for a "safe" level of any chemical not naturally present in the human fetus. The consequence is that whenever we find clear evidence of damage to a fetus related to some blood or bone or tissue concentration of a toxin, it's most prudent to assume at least some damage at all lower concentrations. The alternative attitude, that if we don't see an effect there is no effect, makes no sense with a chemical you already know is toxic at certain concentrations.

The developing brain poses a special problem for toxicology, because the brain is a tissue whose function depends on detailed microscopic organization rather than on the mere bulk of the tissue. Some brain nerve cells and their connections are more important than others for both function and developing organization. Unfortunately, it's not clear where, how, and when developing functionally important organization can be disrupted by a local infiltration of an undetectable concentration of a neurotoxin. What we do understand is that any developing system is likely to be more vulnerable to damage than the same system when fully developed.

OUTSIDE THE FETAL–MATERNAL ENVIRONMENT: INDUSTRIAL POLLUTION

Of toxins in low or high concentrations, we certainly have enough to make us uncomfortable. Chapter 2 described the effects of industry-caused lead poisoning on the developing brain. Now let's look at a few more industrial neurotoxins known or implicated to have disruptive effects on fetal development, particularly development of the brain. The list of pollutants discussed here is far from exhaustive, but serves to illustrate the level of danger that such toxins pose for a vulnerable fetus.

Mercury: A Few Drops to Hell

Located on the west coast of Japan, Minamata City is officially 120 years old, but has cultural roots going back to the seventeeth century, the beginning of Japan's Edo or Tokugawa Period. It started out as a farming village surrounded by a few salt mines and coal mines and became a source of saffron, wax trees, and homemade buckwheat noodle soup. A fishing trade also thrived there, with fish providing the major source of protein for the town. Growing to a population of 30,000 people, the town entered the modern age in the early twentieth century when a large chemical company built a factory there to produce organic chemicals.

The people of the town desperately wanted the factory because they were in hard times. Fishing was still a viable trade, but salt had been nationalized, hydroelectric power was supplanting coal, and the local economy was in decline. They offered the chemical company inducements to build the factory in Minamata: no taxes, the use of land, the use of the port and its facilities, and payment for the power line from the local power plant to the new factory.

The future looked promising. By the late 1930s, the new Minamata chemical plant was a major producer of organic chemicals, particularly acetaldehyde, ethyl acetate, cellulose acetate, vinyl acetylene, acetone, butanol, and isooctane. The town prospered and became a bustling place, housing chemical engineers, chemical workers, and a local service economy for the plant and its employees.

A generation later, in 1956,[30] four cases of what was ultimately recognized as methylmercury poisoning appeared among fishermen who subsequently died. Examination of other death records revealed a total of 17 deaths resulting from the same symptoms: severe convulsions, intermittent loss of consciousness, repeated lapses into psychotic mental states, and a final permanent coma.[31]

The source of the contamination? The chemical plant in Minamata had been dumping mercury, used in the synthesis of acetaldehyde, into Minamata's bay for 24 years. Methylmercury (MeHg) is a combination of a methyl group and the element mercury. The methyl group is a simple organic cluster of one carbon atom and three hydrogen atoms that increases the interaction of mercury with biological substrates. It's the mercury that does the damage to biological systems. In aquatic environments, mercury is methylated by the action of common bacteria, and methylmercury then passes up the food chain and becomes concentrated in sea mammals and in fish, the very staple food on which the people of Minamata and others along the coast subsisted.

"Minamata disease" became a scourge of the west coast of Japan. The chemical plant was shut down, but as of 2001, there were 2265 victims of Minamata methylmercury poisoning, including 1784 deaths. A similar epidemic occurred in the 1960s in Niigata, Japan, also as a consequence of industrial waste discharge.

The dramatic effects of highly toxic chemicals, such as mercury, can often be puzzling to people, since most of us never come into contact with such chemicals. Perhaps this unhappy story of an American incident will help convey just how dangerous such chemicals can be.[32] On August 14, 1996, Karen E. Wetterhahn, age 48, a professor of chemistry at Dartmouth College and a noted researcher on the effects of heavy metals in biological systems, accidentally spilled a few drops of dimethylmercury on her latex glove-covered hand. Dimethylmercury is mercury with two attached methyl groups rather than the single attached methyl group in methylmercury. It is lethal at a dose of approximately 400 milligrams of mercury (equivalent to a few drops), or approximately 5 milligrams per kilogram of body weight. It's classified as a "supertoxic" chemical. Wetterhahn's medical case report makes the following additional points:

1. Records suggest that Wetterhahn handled dimethylmercury on only one day, while wearing the latex gloves and working under a ventilated hood designed to prevent exposure to

chemical fumes. She recalled that the spill occurred while she was transferring the liquid chemical from a container to a capillary tube, after which she cleaned up the spill and then removed her protective gloves.

2. Five months after the accident, on January 20, 1997, Wetterhahn was admitted to the university medical center with a five day history of progressive deterioration in balance, gait, and speech. She had lost 15 pounds over a period of 2 months, and had experienced several brief episodes of nausea, diarrhea, and abdominal discomfort.

3. On February 6th, 22 days after the first neurologic symptoms developed (and 176 days after exposure), Wetterhahn became unresponsive to all visual, verbal, and light-touch stimuli. She died on June 8, 1998, 298 days after exposure.

Following are some anatomical findings from Wetterhahn's autopsy report in the original medical language:

> The cortex of the cerebral hemispheres was diffusely thinned. The visual cortex around the calcarine fissure was grossly gliotic, as was the superior surface of the superior temporal gyri. The cerebellum showed diffuse atrophy of both vermal and hemispheric folia. Microscopical study showed extensive neuronal loss and gliosis bilaterally within the primary visual and auditory cortices, with milder loss of neurons and gliosis in the motor and sensory cortices. There was widespread loss of cerebellar granular-cell neurons, Purkinje cells, and basket-cell neurons, with evidence of loss of parallel fibers in the

molecular layer . . . An extensive high mercury content was
found in the frontal lobes and visual cortex, liver and kidney
cortex. The mercury content of the brain was approximately
6 times that of whole blood at the time of death . . .

You don't need to be a physician or have extensive medical
knowledge to understand what happened to Karen Wetterhahn's
brain as a result of this accident. Her brain was destroyed.

Not every organic mercury compound is as toxic as dimethyl-
and methylmercury. Both compounds are especially powerful
neurotoxins. And the widespread distribution of methylmercury, passing as it does into the food chain after release into the
environment by industry, makes it a pervasive danger to entire
populations of people, including to the unborn.

Indeed, one of the repercussions of the Minamata disaster became obvious in the 1960s, when numerous children were
born with Minamata disease, usually in the form of severe cerebral palsy, low birth weight, microcephaly, profound developmental delay, deafness, blindness, and seizures. Of interest, their
mothers showed no symptoms of methylmercury poisoning and
no methylmercury was detected in their blood. Only later was
it discovered that during prenatal development the placenta
sequesters any methylmercury in maternal blood and passes it
directly to the fetus. The astounding fact is that a developing
fetus can have a high concentration of methylmercury without
it being detectable in maternal blood. Thus, one of the lessons
of Minamata is that a mother can be more or less unharmed by

environmental pollution but still give birth to a child with serious fetal poisoning.[33]

It is of course also possible for both mother and child to
be harmed. An episode of large-scale mercury poisoning more
severe than the Minamata incident occurred in the winter of
1970–1971 in Iraq.[34] In the midst of a famine, 90,000 tons of
seed grain intended for use only as seeds arrived in Basra. The
seed grain was treated with a methylmercury fungicide to prevent rot. The sacks of seed grain were marked as poison, but
only in English and Spanish. Before the sacks of seed grain got
off the dock for distribution to farmers, a large quantity was
stolen and sold to the public as food grain to bake bread. After
consuming the contaminated bread, 6530 people were hospitalized and 459 of these died. Approximately half of the poisoned people were women and many of these were pregnant.
In contrast to the Minamata mothers, the Iraqi mothers started
experiencing neurological symptoms 1 or 2 months after the
poisoning. According to an autopsy report of two Iraqi children exposed prenatally to methylmercury during this incident,
both mothers had above normal methylmercury blood levels,
and both children, who died early in infancy, had above normal
methylmercury levels in brain tissue; their brains were smaller
than normal, showing marked histological abnormalities in the
cerebral cortex and cerebellum.[35]

The Iraqi incident was an acute episode of methylmercury
poisoning from contaminated wheat seeds, and such an episode
is unlikely to occur in an industrialized country. In contrast, the

two Japanese incidents were chronic episodes in an industrialized country, resulting from consumption of fish contaminated by the natural food chain, which in turn had been contaminated by industrial waste. Contamination of food fish by industrial-waste methylmercury continues to be a problem anywhere people eat fish, and methylmercury has become as important as lead as a prenatal neurotoxin.

The control of mercury pollution is often influenced by politics. Throughout the 1990s, the Environmental Protection Agency (EPA) made progress in reducing industrial mercury emissions, but under the administration of George W. Bush, the policy changed. In 2003, for example, the EPA proposed to reverse strict controls on emissions of mercury from coal-fired power plants. A Clear Skies Act, proposed in 2003 by Republicans to amend the Clean Air Act first passed in the 1960s, is stalled in Congress and has been criticized by pediatricians as an effective substantial relaxation of pollution control.[36]

Meanwhile, approximately 500,000 American children are born each year with umbilical cord mercury blood levels great enough to suffer mercury-related losses of cognitive function ranging from less than 1 point to as high as 24 IQ points.[37] An IQ loss of 24 points is not trivial: It's enough to render a child borderline mentally retarded. In fact the downward shifts in IQ in America resulting from prenatal exposure to methylmercury of anthropogenic origin are estimated to be associated with an average of 1566 excess cases a year of mental retardation.[38]

In general, the existing evidence from a range of studies shows an average reduction in offspring of 0.2 IQ points for each part per million increase of maternal hair mercury.[39]

Arsenic

A natural constituent of the planet, arsenic is found everywhere in bedrock, and it easily dissolves into ground water to eventually contaminate drinking water. Cross-sectional studies of school-age children have shown cognitive deficits associated with arsenic in drinking water coupled with raised arsenic concentrations in urine. Similar observations have been made of children exposed to arsenic from a smelter. Although evidence for subclinical neurodevelopmental neurotoxicity of arsenic is less-well established than for lead and methylmercury, the data that do exist are consistent and fit with the high-exposure findings.[40]

Dioxins and Polychlorinated Biphenyls

Part of the problem of environmental pollution is that many new chemicals, once in the environment, are almost impossible to remove from air, water, soil, and living systems. Such is the case for many dioxins and polychlorinated biphenyls (PCBs)—new chemicals that were hailed when they arrived and then castigated when we found them to be dangerous environmental poisons.

The term "dioxin" refers to a class of heterocyclic hydrocarbons compounds used in various industrial processes, such as paper manufacture and waste incineration. They readily pass into and accumulate in living systems. Animal experiments have demonstrated that these chemicals can produce mutations and fetal malformations. They have also been implicated as human carcinogens.

PCBs are another group of chemicals with many industrial applications (for example, adhesives, wood floor finishes, and paints), and like dioxins are extremely stable and persistent organic pollutants. In the United States, they were used commercially for 50 years before they were banned in 1979, and with good reason: It became evident that PCBs can cause cancer, and that they can have a variety of bad effects on the immune, reproductive, nervous, and endocrine systems. They accumulate in high concentrations in fatty tissues, with concentrations magnified up the food chain. They are present in common foods, such as meat, fish, eggs, and milk, and are still found as contaminants in air, water, and soil, and in many old products manufactured before 1979, and can still be detected in the blood of nearly all Americans (and in most people in other industrialized countries). One of the most well-known sites of PCB pollution is New York's Hudson River.[41] Contaminated for decades with toxic waste from power plants along its banks (including an estimated 150,000 pounds of PCBs), it is the longest Superfund Hazardous Waste Site in the United States. Nearly all fish in the Hudson River are contaminated, and nearly all commercial fishing in the river has been banned.

Researchers argue about the effects of prenatal exposure to PCBs on brain development and later childhood cognitive performance. Some studies find effects and some don't. Some studies do not involve direct measurements of PCBs in blood but rely on surrogate (proxy) measurements, such as consumption of fish from contaminated waters. It's also difficult to locate a population in which fetuses have been exposed to PCBs but not to other toxins. It may be that cognitive effects are limited to certain kinds of brain-work not revealed by current tests. We don't know yet. But given all the evidence that exists about serious effects in animals and human adults, and given the general axiom that the human fetus is about a hundred times more vulnerable than adults to toxins, it's reasonable to say that the question of the effects of prenatal exposure to PCBs is not yet resolved.

Meanwhile, what do the studies that point to these effects specifically indicate? Some examples are as follows

- In 1996, researchers reported intellectual impairment (in short-term memory, long-term memory, and sustained attention) in school children exposed to PCBs in utero.[42] The data showed that in the prenatal environment the presence of PCBs in concentrations only slightly higher than those in the general environment can have a long-term impact on intellectual function. In addition, the deficits in short-term memory in infancy and early childhood are consistent with reports of reduced IQ scores of children in Taiwan whose mothers had ingested rice oil contaminated with PCBs and dibenzofurans (a class of organic compounds with many uses, including as insecticides and industrial bleaching agents).

- In populations that consume large amounts of Great Lakes fish contaminated with PCBs, PCBs may be important contributors to subtle neurobehavioral changes in newborn children.[43] Some of these changes persist during childhood.
- PCB concentrations in umbilical cord blood at birth have been related to later cognitive deficits in children at 8 years of age.[44]
- PCB exposure is apparently correlated with socioeconomic status. The blood PCB level of inner-city African-American women correlates positively with income[45]—the higher the income, the more PCBs in their blood. Similar results have been found in the American white population, an apparent reflection of differences in housing and diet. PCBs are one instance in which effects on the middle class are apparently greater than effects on people in poverty.

Endocrine Disruptors

Endocrine disruptors are exogenous chemicals (for example, PCBs) that enter the body and act like hormones. They can disrupt physiology, sometimes with devastating consequences. Startling evidence from animal studies shows that male fish in detergent-contaminated water express female characteristics, turtles are sex-reversed by PCBs, male frogs exposed to a common herbicide form multiple ovaries, pseudohermaphroditic offspring are produced by polar bears found in contaminated waters, and seals in contaminated water have an excess of uterine fibroids.[46]

In addition, transient exposure of a pregnant female rat during the period of gonadal sex determination to the endocrine disruptors vinclozalin (an antiandrogenic compound) or methoxychlor (an estrogenic compound) induces decreased spermatogenic capacity (cell number and viability) and increases incidence of male infertility in nearly all males of all subsequent generations examined.[47] These effects on reproduction correlate with altered DNA methylation patterns in the germ line. An endocrine disruptor can apparently reprogram the germ line and promote a transgenerational disease state.

Can we expect environmental endocrine disruptors to affect the sexual development of the human fetus? In my opinion, the answer is an absolute yes, if only because we know that endocrine disruptors act like foreign hormones and thereby disrupt natural hormone physiology. Hormones are critical in fetal sexual and brain development. Exposure of the fetus to hormones can produce profound changes in development. In the case of androgenic hormones, for example, exposure diverts a genetic female to take on the phenotypic appearance of a male, and these hormones change the areas of the brain that ordinarily differ between the sexes (sexually dimorphic areas).

Air Pollutants

Air pollution is a serious problem in almost all large cities in the world. Combustion emissions account for over half the fine-particle content and particulate organic matter in urban polluted air. The mutagenic and carcinogenic constituents have been studied and measured in populations in America, Europe, Asia, and many developing countries. Moreover, evidence has accumulated that indicates air pollution can have a destructive impact on fetal development. Animal and human data show that polycyclic aromatic hydrocarbons, common constituents of air pollution, can cross the placenta and reach fetal organs, and animal data suggest involvement of these chemicals in various kinds of fetal damage.[48] The fetus is apparently especially sensitive to DNA damage from air pollution and such damage can induce fetal somatic mutations. In 2002, researchers reported that environmental pollution involving aromatic DNA-adducts (compounds that complex with DNA), such as that from the combustion of fossil fuels, can produce in utero DNA damage in the fetus and associated somatic gene mutations in newborns.[49] DNA-adducts almost certainly will also affect gene expression—and often with subtle subclinical consequences.

In general, exposure of pregnant women to air pollution is associated with adverse fetal and childhood outcomes, such as low birth weight; preterm birth; intrauterine growth restriction; congenital defects; intrauterine and infant mortality; decreased lung growth; increased rate of respiratory tract infections; childhood asthma; behavioral problems; and neurocognitive deficits.[50]

It should be obvious from this chapter that the developing human fetus is in no way insulated from the outside world.

Prenatal development is subject to a variety of environmental impacts that can divert, disrupt, and stop development altogether. The effects may not be observable at birth, but many of them affect later cognitive performance and emotional stability that may remain subclinical and may never be considered as deriving from the fetal period.

Reviewing again the Barker hypothesis might help to drive this point home. As noted previously, the central feature of the Barker hypothesis is the idea that the human fetus can respond to environmental effects—in this case, unbalanced nutrition—by permanently changing its developmental and growth trajectories. Put another way, the hypothesis proposes that maternal under-nutrition acts during pregnancy to program risks in the fetus for adverse health outcomes, such as cardiovascular disease; obesity; and metabolic syndrome in adult life. An amplification of this idea is that fetal programming is an evolutionary adaptation that activates the hypothalamic–pituitary axis to produces an early fitness advantage (fetal survival) at the cost of later adult disease.

This raises an important question: Is it possible that evolved fetal programming to achieve fetal survival can also occur at the cost of later dysfunctions of brain and behavior? We don't yet know the answer to this question. What we do know is that the neurological and behavioral consequences of various environmental effects on the fetus range from subtle to catastrophic. The evidence concerning fetal programming that results in physical

diseases is an important new reason why any attempt to understand the origins of behavior must not ignore the fetal environment and its cluster of variables. In the following chapters, I'll
discuss details about specific behavioral consequences, including
some that may be subclinical, of various environmental impacts
on fetal development.

PART II

THE FORGE OF CIRCUMSTANCE

SIX

The Endless Fetal Hangover

Claire is 26-years-old and married to a young man she has known since high school. His name is Gordon and he's an attorney. Claire teaches third grade in a private elementary school in Manhattan and she loves her job. She loves children. She wants children of her own. It's now 10 o'clock in the evening of Gordon's birthday, and the couple is at a party on the Upper West Side, 12 young people laughing, talking, teasing each other, rolling their eyes whenever someone tries to be silly. No one is smoking. This crowd doesn't smoke anything. But they drink this or that, good wine, vodka over ice, spiked punch, and Claire and Gordon are drinking more than usual. Before long Claire is unsteady on her feet and she needs to sit down. But she keeps drinking, her drink of choice a cool vodka tonic.

How many has she had by now? Five? Six? She doesn't remember anymore. The vodka tonic is delicious. The more she drinks, the better it tastes. So Claire drinks. Life has been good to her. She's 26 and carefree. More vodka? Gordon asks. He brings her another glass. It's a wonderful evening, isn't it?

Maybe not. Although she doesn't know it yet, Claire is pregnant. Some women can binge drink during early pregnancy and nothing much happens to the child they carry. Claire isn't one of those women. On the night of the party, her embryo is just beginning the second week of gestation, and the exposure to alcohol during this one evening is the beginning of a human tragedy. Three weeks later, Claire stops drinking when she learns she's pregnant. But the damage to the fetus has already occurred. The embryo in her uterus was vulnerable to a sudden increase in blood alcohol. Or maybe Claire's body detoxifies alcohol too slowly. We don't know the mechanisms for individual vulnerabilities. All we know is that soon after Claire's son is born several months later, he has physical problems, mental problems, and an unpredictable future. Claire's child, the child she yearned to have, will be diagnosed with fetal alcohol syndrome (FAS), including mental deficits and a related cardiac defect.

The most insidious fetal neurotoxin known to us is available as a recreational product in every supermarket in cans and bottles large and small. It is ethyl alcohol (ethanol). In the brain, it acts as a depressant, reducing the activity of nerve cells by acting on neuron membranes to prevent nerve impulses. When it appears to stimulate adult behavior, it's the result of depression of inhibitory neural circuits.

Because of its chemical solubility properties, alcohol easily penetrates all cells and tissues in the body. It's a neurotoxin for the central and peripheral nervous systems, a hepatotoxin for the liver, a nephrotoxin for the kidney. It can interact chemically with any biological cell, including a progenitor cell in an embryo; it can kill that cell and the entire lineage that would develop from it during embryonic development.

Approximately half of American children born each year are exposed to ethyl alcohol during fetal development. As a fetal teratogen and neurotoxin, it's a serious public-health problem—the cause of more mental retardation in America than all other known causes combined, including chromosomal defects. It is also the cause of more subtle cognitive effects: Children exposed to alcohol in utero may have intelligence quotients (IQs) in the normal range but later show deficits in cognitive skills and language development.

And the exposure is often unknown to mothers who, like Claire, drink while unaware that they are pregnant. That unawareness, coupled with high fetal vulnerability to ethyl alcohol, is one reason the substance is so insidious. Its often celebrated place in the public mind only exacerbates the problem.

THE HUMAN LOVE AFFAIR WITH ALCOHOL

Booze, blotto, bottoms up—the slang words in English for alcohol and drinking go back to the thirteenth century. But alcoholic beverages have a much longer history than that, one that apparently goes back at least 12,000 years. Distillation was brought to Europe by the Arabs in the Middle Ages. The word "whiskey" is Gaelic for "water of life"—an elixir for all diseases.[1]

This heritage suggests that we have always loved our alcoholic beverages. True, in the United States during the early twentieth century, the temperance people hated alcohol, said it did the devil's work, but just about everybody else loved booze in any form and had a grand time getting drunk. Prohibition of alcohol, from 1920 to 1933, made piles of money for bootleggers and owners of speakeasies, but hardly reduced public consumption.

A year after Prohibition ended, alcohol began to gain a special glamour with the six "Thin Man" detective films that came out of Hollywood between 1934 and 1947. The films starred William Powell and Myrna Loy, with Powell playing a debonair Manhattan private detective (Nick Charles) who moved with his wife Loy (Nora Charles) from bar to nightclub to bar guzzling martinis one after the other. Hollywood gave Americans what they wanted: an alcoholic vision of a good life in which knowing how to shake a martini was a sign of good breeding.[2]

No one knew enough about alcohol in those days. People thought that as long as you controlled your drinking without actually becoming an alcoholic, no harm was done. So everybody drank, including pregnant women. The harm? Using current prevalence figures, we can estimate that in the 1930s some 25,000 newborn babies each year in America were born mentally retarded because their mothers drank alcohol during gestation. But no one apparently made the connection. Not until the 1970s, as we'll discuss shortly, was the link between fetal exposure to alcohol and developmental damage identified. Eventually, it was also understood that even in small concentrations ethanol is a potent developmental neurotoxin.

ETHANOL AND EVOLUTION

The human love affair with alcohol is sometimes thought to be a Darwinian evolved adaptation. As I mentioned in the previous chapter, ethanol is a naturally occurring product of fermentation of fruit sugars by yeast. Given that, some researchers have suggested that positive behavioral responses to ethanol may have been the target for natural selection for all fruit-eating (frugivorous) species, including primates, hominoids, and our own hominid line.[3] Consumption of low-concentration ethanol within fruit may act as a feeding stimulant. The scent of ethanol might have been used by animals, including early humans, to locate ripe fruit. It's a reasonable idea, even if it leads to arguments about adaptation. Moreover, the alcohol in ripe fruits may not be harmless. Some ripe fruits can have an alcohol concentration as high as 4.5 percent, probably enough to affect the developing brain of a fetus if such fruits are consumed by pregnant women.

Is consumption of alcohol an evolutionary adaptation? Darwinian natural selection is for the most part blind to behavioral traits that appear only after the human reproductive age—such traits cannot change the probability of reproduction. Maybe natural selection is also blind to brain damage that does not interfere with reproductive fitness. But the idea that every common human behavior trait, such as alcohol consumption, has an adaptive purpose is a media myth. Some traits may have been adaptive in the past, but new environmental conditions can make a trait maladaptive or even lethal. And some traits have never had an adaptive purpose—they appeared as "spandrels"—evolutionary free-riders on the backs of other traits that are adaptive in the short or long term.[4] The key is that when Darwinian natural selection favors an adaptive trait, the gene complex responsible for that trait may also be responsible for other traits that are spandrels—traits with little or no positive effect on survival to reproduction.

But no matter how or why such a trait as the human taste for alcohol evolved, people do drink alcohol and pregnant women drink alcohol before and after they know they're pregnant. So what, in detail, are the consequences of this behavior on the children these women bring into the world?

FETAL ALCOHOL CONDITIONS

The consequences of fetal alcohol exposure range from subtle to serious damage. The term "fetal alcohol spectrum disorder" (FASD) refers to any alcohol-related defects. The less severe cases in children are identified as "alcohol-related neurodevelopmental disorder" (ARND); it is, however, often difficult to recognize this condition at any age, resulting in an underestimate of prevalence.[5] At the most severe end of the spectrum is what Claire's son has: "fetal alcohol syndrome," which includes fetal structural defects, fetal growth deficiency, and mental retardation. It is easiest to recognize in children 2 through 11 years of age, but more difficult to recognize in newborns unless the bodily changes (dysmorphology) are obvious,[6] including recognizable facial anomalies: short eyelid fissures, a thin upper lip, and an absent philtrum (the groove between the middle of the upper lip and the nose). Some FAS children also have other anomalies such as dropped eyelids (ptosis), a flat midface, an upturned nose with a flat nasal bridge, underdeveloped ears, and cardiac defects. Prenatal or postnatal growth retardation is also common.[7]

FAS was first identified in 1973, in two historic papers appearing in the British medical journal *The Lancet*. The first paper opened as follows[8]:

> Eight unrelated children of three different ethnic groups, all born to mothers who were chronic alcoholics, have a similar pattern of craniofacial, limb, and cardiovascular defects associated with prenatal-onset growth deficiency and developmental delay. This seems to be the first reported association between maternal alcoholism and aberrant morphogenesis in the offspring.

The second paper, by the first two authors of the first paper, named the condition "fetal alcohol syndrome" and added three more cases.[9] Its short historical review pointed out that "the association between maternal alcoholism and faulty development of the offspring is alluded to in early Greek and Roman mythology."

Indeed, alcoholism among both men and women is not a new phenomenon, and certainly not in the Western world. Pregnant women who drank alcohol were always visible everywhere, known to everyone including physicians. Women have been drinking wine at dinner tables since dinner tables were invented. The birth outcomes of alcoholic women were also known to physicians. In 1834, a report in the British House of Commons by a committee investigating drunkenness noted that infants born to alcoholic mothers sometimes had "a starved, shriveled and imperfect look."[10]

But after that report interest apparently withered. From the beginning of modern medicine in the early nineteenth century to 1973, nothing but sporadic clinical reports appeared suggesting a link between maternal alcoholism and fetal damage. With the *Lancet* papers, interest in the connection increased, such that the existence of FAS began to raise questions about the role that *any* degree of alcohol consumption might play in fetal abnormalities.

Simply put, it's not necessary for a pregnant woman to be an alcoholic to produce FAS or FASD or ARND in a fetus. It's also not necessary for such a woman to have any of the vulnerabilities of pregnant women in poverty. The fetuses of middle- and upper-class pregnant women are no more immune to alcohol-related abnormalities than the fetuses of their poorer counterparts. In general, from the standpoint of public health, the most important datum is whether there is *any* drinking during pregnancy, not how much, not whether so-called binge or nonbinge "social" drinking occurs. The reasons for this are clear: it's apparent that there is no threshold below which alcohol use during pregnancy is safe: Its use even at low levels during pregnancy has harmful effects on later child behavior.[11] The effects are observed at average levels of alcohol exposure as low as one drink per week. Furthermore, research with animals gives us startling data about the toxicity of ethanol. A single exposure of infant rats or mice during the formation of neural connections (synaptogenesis, which occurs in the mid to late pregnancy stage for humans) can cause developing neurons to self-destruct (apoptosis) on a massive scale.[12] The corresponding human dose during pregnancy that would lead to loss of fetal brain cells would be two drinks.[13] One drink is 12 ounces of beer, 5 ounces of wine, or 1.5 ounces of "hard" liquor.

Alcohol is highly toxic in the fetuses of pregnant women because its high solubility in both oil and water enables it to easily cross the placenta. The level of alcohol in fetal blood rises almost as rapidly as the level of alcohol in maternal blood. Within an hour, the level of alcohol is the same in both mother and fetus.[14] Approximately 12 hours are needed for alcohol to be cleared out of maternal and fetal blood. Animal studies suggest, however, that significant alcohol may remain in amniotic fluid after blood clearance. If alcohol is in amniotic fluid, it means the embryo and fetus are exposed to its effects on developing tissues in the brain and elsewhere.

The problem of alcohol use before and during pregnancy has at least been officially recognized by the federal government. On February 21, 2005, the Surgeon General of the United States released this unequivocal advisory[15]:

1. No amount of alcohol consumption can be considered safe during pregnancy.
2. Alcohol can damage a fetus at any stage during pregnancy.
3. The cognitive deficits and behavioral problems resulting from prenatal alcohol exposure are lifelong.
4. Alcohol-related birth defects are completely preventable.
5. A woman should not drink alcohol during pregnancy.
6. A woman who is considering becoming pregnant should abstain from alcohol.

Unfortunately, federal billboards and TV advertising emphasizing the points of the advisory are hardly in evidence. Meanwhile, industry expenditure on alcohol advertising in America is approximately $2 billion a year. At the same time, according to current epidemiological estimates, 10–25 percent of pregnant women are drinking alcohol during their pregnancy and giving birth each year to approximately 40,000 children with varying degrees of brain damage.[16] Maybe double that percentage are women who don't know they have conceived and who drink alcohol ranging from moderate to binge levels.

Various estimates amplify the problem (see also later sections):

In the United States, 45 percent of all women consume alcohol during the 3 months before they learn they're pregnant.[17] Sixty percent of women who consume alcohol also do not learn they're pregnant until after the 4th week of gestation, and many do not know until after the 6th week. We can assume these women are drinking as much alcohol during the first 4–6 weeks after conception as they were drinking before conception.

Of course, in a country as large and as diverse as America, prenatal alcohol exposure varies from place to place. Of pregnant low-income women in Southern California (Los Angeles and Orange counties), 24 percent consume alcohol past conception.[18] Of that percentage, 67 percent drank before pregnancy recognition, and 33 percent continue to drink after pregnancy is confirmed.

Approximately 30 percent of white non-Hispanic, black non-Hispanic, and English-speaking Hispanic women drink past conception, compared with 16 percent of Spanish-speaking Hispanic women. This supports the idea that acculturated Hispanic women (English-speaking) tend to incorporate the drinking patterns of the larger U.S. population to a greater extent than less acculturated Hispanic women. Available warning messages and the level of knowledge among Hispanic-American women about alcohol-related fetal abnormalities are evidently not enough to prevent early pregnancy alcohol consumption, a situation apparently typical of low-income ethnic groups in America.[19]

Of women aged 18–44 years, approximately 2 percent of pregnant women and 13 percent of nonpregnant women engage in binge drinking.[20] Approximately 28 percent of these women consume an average of 5 drinks or more on typical drinking days, with 21 percent of these women consuming at least 45 drinks on average in a month. Larger proportions of binge drinkers are found among women of younger ages (18–24 years) or current smokers.

A serious problem is the concurrent use of both alcohol and tobacco by young women who are pregnant but don't know it yet. The potential to expose an embryo to both tobacco and alcohol is high among these women. Some researchers are pessimistic about reducing the prevalence of concurrent alcohol and tobacco use in this group, noting that addictions to nicotine and alcohol are progressive and chronic.[21]

And the prevalence of alcohol-related fetal abnormalities? The incidence of FASD in the United States is currently approximately 1 per 100 live births.[22] This means that this disorder may affect about 1 percent (3 million children and adults) of the total American population and occurs more often than Down syndrome and spina bifida combined. The U.S. prevalence rate of the more severe form, FAS, is much less, ranging from about 5 to 50 per 10,000 live births, depending on geographic location and ethnic group.[23] Blacks and American Indians have the highest prevalence of FAS. But the most serious effects appear to occur especially in African-American children exposed to alcohol during fetal development.[24]

NEUROLOGICAL AND BEHAVIORAL EFFECTS OF FETAL ALCOHOL EXPOSURE

Children with FAS usually struggle in school because of decreased cognitive functioning and social problems, but diagnosis of their condition is often delayed or missed entirely. Children with cognitive deficits from FASD usually do not manifest any of the dysmorphologies of FAS. A troubled child with FASD usually does not look any different than any unaffected child. Nonetheless, we do know something about the neurological and behavioral difficulties that children with FAS, FASD, or related effects of prenatal alcohol exposure generally present.

So-called executive brain function, examined and scored by psychological testing, is defined as higher order psychological processes involved in goal-oriented behavior under conscious control. In general, it's an umbrella term for many cognitive processes that include planning, organized searching, inhibition, working memory, set shifting,[25] flexible thinking, strategy employment, and linguistic fluency.[26] This complex of behaviors is believed to be mediated by the frontal lobes of the brain.

Executive function deficits are common among people with FASD, and such deficits are related to crime. Sixty percent of FASD individuals have a history of trouble with the law and 50 percent have been in jail or prison at one time or another. Executive function deficits are also common in other disorders, such as attention deficit hyperactivity disorder, autism, various learning disabilities, and traumatic brain injury.[27]

Children at age 7 years who are prenatally exposed to maternal binge drinking are twice as likely to have IQ scores in the mentally retarded range, and 2.5 times more likely to have clinically significant levels of delinquent behavior.[28]

What about the effects on IQ of moderate drinking during pregnancy? If a pregnant woman consumes on average two or more drinks a day during pregnancy, the reduction in IQ of her child at 7.5 years of age is about 7 IQ points.[29] Examination of infants 12 to 13 months old with a standard infant cognitive performance test (Bayley Mental Development Index) shows that maternal drinking of even less than one drink per day during pregnancy significantly reduces her child's cognitive performance.[30] One study found that heavy prenatal alcohol exposure, with or without FAS, is associated with social conduct disorders, particularly with juvenile delinquency, in children 10–18 years of age.[31] In general, subtle destructive effects of prenatal exposure to alcohol can affect later academic and social functioning even with maternal alcohol consumption at social drinking levels. This is now the accepted view in the American medical community.[32]

Prenatal exposure to alcohol is also a predictor of alcohol abuse later in life.[33] Moderate consumption of alcohol by a mother during pregnancy is associated with increased responses of infants to the odor of alcohol even if the infants show no signs of FASD.

New techniques are being used to locate brain damage from fetal alcohol exposure. Magnetic resonance imaging (MRI), for example, has been used to analyze the shape of the corpus callosum (the thick band of nerve fibers connecting the two hemispheres of the brain) in adult males who were exposed prenatally to alcohol and relate the measurements to neuropsychological deficits.[34] The men showed more variability of corpus-callosum shape and the excess shape variation is associated with behavioral deficits. A relatively thick corpus callosum is associated with deficits in executive function, while a relatively thin corpus callosum is associated with deficits in motor function.

MRI studies of children with FAS also show reduced volume of other specific brain structures, the reduced volume not related to any reduction in brain size.[35] The caudate nucleus is an example. Damage is evident in the caudate nucleus of a variety of children with FAS and lesser fetal alcohol effects.[36]

Experiments that reveal the effects of alcohol on nerve cells of the prenatal developing brain in humans are not feasible, but animal experiments show that exposure of the fetus to a relatively low level of ethyl alcohol in maternal and fetal blood promotes significant changes in the migration of neurons to and in the cerebral cortex.[37] The wiring of the prenatal brain is affected by alcohol exposure.

Eyeblink conditioning is a Pavlovian reflex developed after pairing a conditioned stimulus (for example, a beep) with an unconditioned stimulus that always causes an eyeblink (for example, an air puff to the eye). Among 5-year-old children, no child with FAS is able to develop this reflex, as opposed to 87 percent of control children.[38] These results persist when controlling for IQ. Most children with abnormal smallness of the head (microcephaly) but without FAS do establish the reflex. Animal studies have shown that alcohol consumption during pregnancy also impairs the eyeblink reflex in animal offspring. The cerebellum, a brain control center for the eyeblink reflex, is evidently involved. The fetal alcohol exposure impairment apparently involves a loss of neurons in the fetal cerebral cortex and cerebellum and a loss of plasticity in the cerebellum.

Each year in America, the national consumption of alcohol by women of child-bearing age adds approximately 100,000 more mentally retarded children to the population. How many more children not diagnosed as mentally retarded are affected by prenatal exposure to alcohol? We know only that this exposure is related to atypical brain development, neuropsychological deficits, academic difficulties, emotional dysfunction, and social dysfunction.[39] Is this constellation of problems acceptable collateral damage of the alcoholic beverages industry?

Meanwhile, the high prevalence of alcohol use by pregnant women and women of child-bearing age, coupled with the high prevalence of FASD, makes at least one idea seem bizarre: the idea that the fetal environment contributes little or nothing to postnatal behavior and IQ.

SEVEN

Unborn Days and Sexuality

On a spring day in Minneapolis, a tall attractive woman saunters down a quiet residential street. Most people are at work or at school, and she encounters no one as she strolls on the clean sidewalk between a row of houses and a row of parked cars at the curb. Call her Marcia. She's carefully groomed and well-dressed, maybe too well-dressed for this quiet middle-class neighborhood at this time of the day. She wears a red hat with a wide brim, a flowing red dress, beige stockings, and red high-heeled shoes. Her walk is graceful and confident; the walk of a woman who knows that she's sexually attractive. She has flowing auburn hair, a lovely face with classic features, a pretty mouth painted with lip rouge whose color exactly matches the color of her dress. Her dress is tight enough at the bodice and hips to suggest female curves, what the French call "une belle tournure."

Apart from the fact that she's attractive and she may be overdressed, nothing seems special about Marcia. But the reality is she has a secret, a secret covered by her clothes. At the joining of her thighs, Marcia has the genitals of a male. She's a transsexual, an individual born with the genitals of one sex and the psychological attitudes of the other sex.

Marcia's feminine appearance has been enhanced by hormone injections. She has the breasts and buttocks and general softness of a woman, but she still retains complete male genitals. She's not hermaphroditic—she has the organs of only one sex. She's also not a simple male transvestite—she's not a man who merely dresses as a woman.

Marcia has no confusion about her identity. She looks at herself as a woman; she thinks of herself as a woman, she knows she's a woman. She will soon have "sex reassignment" surgery, a new genital construction, and she will complete her change to a male-to-female transsexual.

Marcia is one of about 10,000 American male-to-female transsexuals.[1] The number of parallel female-to-male transsexuals in the United States is much less, about 3000 in the American population.[2] Not all transsexuals choose to undergo surgery to make a complete genital transition. That's about all the "choice" involved. The mismatch between genital identity and psychological identity was never chosen by Marcia.

So how did Marcia happen? Where did this woman's sexual odyssey begin?

As recently as 30 years ago we had hardly any answers. But in the 1970s, the new biochemistry coupled with the new molecular biology and the new understanding of endocrine effects began pointing to a shocker. It seems that the sexual odyssey of many people, both men and women, begins not only with the genes they've inherited, but also with the influence of their prenatal environment. If the prenatal development of the sexual organization of the brain is pushed one way, you get one result; if it's pushed another way, you get another result. Our psychological sexuality is not simply preordained in our genes, but it's also not a simple result of childhood and adult experience.

We're learning that, as in every other aspect of human ability and behavior, the prenatal environment plays a critical role in the origin of our sexuality and gender identity. Both derive from complex processes that begin at conception and the general programs for these processes have evolved by Darwinian evolution. Put another way, Darwinian evolution has given us the genetic programming for fetal sexual development and gender identity during gestation, but the execution of that programming is always dependent on, shaped by, and potentially altered in subtle or even radical ways by the local environment of the fetus—that is, by the physiology and biochemistry of the mother and by impact from the mother's environment. The Darwinian programming continues after birth, although learning experiences during childhood can further modify the programming.

ANIMAL ODDITIES: SEXUALITY IS NOT SIMPLE

The animal world provides us with a huge zoo of fabulous natural experiments in sexuality. Consider the slipper-snail, *Crepidula fornicata*. It lives in a shell about 2 inches in diameter, and it's a protandrous hermaphrodite, which means males can change into females. At mating time, a large female climbs onto an empty shell, and soon a male climbs on top of her and copulates with her. Before long another male climbs on top of the first male, the first male changes into a female, and the second male copulates with her. A third male then climbs onto the second male, and now the second male changes into a female—and so on, until a chain of maybe a dozen slipper-snails is formed, usually from larger to smaller at the end of the chain, where the final slipper-snail remains a male, copulating with the female below him. What else is this but group sex ordained by Darwinian evolution?

Or have a look at the marine worms called Bonellia, blue-green in color and about 6 inches in length. They were discovered in the Bay of Naples in Italy in 1821, and for a long time the only form known was the female, a lively animal living at depths as much as 350 feet. Where were the males? After much searching, zoologists concluded that maybe there were no males at all. Then finally the males were discovered—living in the oviducts of the females, six or seven dozen tiny males living inside each female. The human equivalent would be a wife with a harem of little husbands carried in her fallopian tubes. We call it polyandry, but Darwinian evolution invented it a long time ago.

Consider any sort of sexuality you can imagine, and it's probable Darwinian evolution has already made it a reality. But what is sexuality? The term "sexuality" is a construct, a mental category convenient for the way the mind works. Every language has many such constructs, but the fact that we can invent and use a construct doesn't mean that what it refers to is necessarily as succinct as the construct itself. We divide sexuality into male and female, a simple and sharp division. But that dichotomy is our problem, not nature's. Nature doesn't speak English—or any other human language. We're not going to understand human sexuality as long as we pretend the constructs currently in use in our language are always isomorphic with reality. As we discover more and more about the real world, some of our pet constructs begin to lose their relevance—and "sexuality" is one of those constructs.

Indeed, constructs can become exceedingly confusing, as manifested in the endless debate about the difference between the terms, *sex* and *gender.* Sociologists use the term "sex" when referring to the biological attributes of genitals and chromosomes—the biological identity. They use the term "gender" when referring to psychosocial identity—how individuals classify themselves, or how others classify individuals: as males, females, or ambivalent. But others often use the two terms ambiguously, illustrating just how muddled we are about sex and gender issues. Throw the term *sexual orientation* into the mix, or bring up the topic of homosexuality versus bisexuality versus heterosexuality, and all manner of debate, much of it confusing, also unfolds without resolve. I will take a stand here and use the term "sex" to refer to biological attributes (classical primary and secondary sexual characteristics) and "gender" to refer to personal identity (psychological or brain gender)—but I will not use the term gender to refer to how others classify an individual, because I don't think that's relevant: there are too many ambiguities that result when we take into account what society thinks when it looks at people. So in what follows we will let the personal identity decide the "gender" of the individual. This means "gender confusion" is personal confusion and not confusion of society looking at that person. And it means homosexuality is not a "lifestyle" but a personal (brain) identity.

Still other constructs have been mired in biological ignorance. Consider, for example, fetal sexual development and certain attitudes about it. The ancient Greeks believed that the sex of the fetus is determined by the "heat" of the male partner during copulation—the more heated the passion, the greater the probability of a male fetus. Aristotle advised men to copulate in the summer if they wanted male heirs. The idea became solidified in Western thought and was the common view for the next thousand years into the Middle Ages.

Another myth derived from the ancient Greeks and Aristotle, a myth that lasted several centuries after the Middle Ages, was the idea that females were simply males whose fetal development had for some reason stopped too early. The female was considered a defective male, full development halted because the mother's womb could not overcome the heat of the male semen. This view was accepted by the European Church after Galen, then the supreme authority in anatomy and physiology, declared the following in the second century[3]:

> Just as mankind is the most perfect of all animals, so within mankind the man is more perfect than the woman, and the reason for this perfection is his excess heat, for heat is Nature's primary instrument ... The woman is less perfect than the man in respect to the generative parts. For the parts were formed within her when she was still a fetus, but could not because of the defect in heat emerge and project on the outside.

How many mothers, during all these centuries, have hinted to their little daughters while bathing them that girls are "missing something down there"? In one form or another, the general idea of female anatomical inadequacy persists as a social undercurrent even today, and until recently we had a public bemused by media psychologists claiming that the personalities of girls and women are molded by "penis envy"—by a genital deficiency.

Beliefs about other factors that supposedly determine sex also persisted until the end of the nineteenth century, with biologists claiming that constitution, age, nutrition, and environment of the parents were especially important in determining the sex of the fetus.[4] The experts argued that factors favoring the storage of energy and nutrients predisposed the birth of human females, while factors favoring the utilization of energy and nutrients produced human males. These ideas may seem ridiculous now, but we need to remember that during the nineteenth century biology was still mainly a descriptive rather than an experimental science and fundamentals that we now take for granted were either unknown or only speculations.

PRENATAL DEVELOPMENT AND SEX HORMONES

We now understand that (genital) sexuality and (brain) gender identity in the fetus are the result of complex interactions between the brain, hormones, and hormonal target tissues, and that after birth, psychological input to the brain from the environment can further twist development in one direction or another. But that understanding has only emerged in the past 20 years, during which we have accrued more knowledge about fetal sexual development than during the past five thousand years of written history.

Certainly, some important findings occurred previously, as in the animal research of the 1940s that first introduced the idea of hormonal regulation of the sexual differentiation of fetal reproductive organs.[5] If testes are removed from a fetal male rabbit as soon as they appear, the fetus develops into a female. If testes are removed from a fetal male rabbit and transplanted into a female, the previously female fetus develops into a male. The only reasonable conclusion is that the early fetus has the potential to be either male or female and the outcome depends on hormones secreted by the fetus itself. In the absence of hormonal control, the default development is that of the female reproductive system.

Animals are not always good models for humans, but in this case what happens in rabbits also happens in our own species. From an embryological standpoint, the penis and clitoris are the same organ, one appearing in men and the other in women. In the embryo, before the genitals fully form, there's a precursor organ that develops into a penis if it's exposed to certain hormones, without which it becomes a clitoris. Does this mean that in principle we can artificially control the genital identity of a developing human fetus? Yes, it's possible in principle. And maybe in a few hundred years or less artificial sex determination, the control of both biological sex and pychosocial gender, will be routine. I'm not proposing this as a positive development—I can't make that value judgment. But I do think that, given the way we live now, if in the near future some people will want artificial sex determination, and some people will be able to make money at it, artificial sex determination will indeed be available.

Meanwhile, the sexual development of the human fetus is shaped by natural hormonal programming, a consequence of genetics, epigenetics, and a fetal environment that is often malevolent.

Even in adults, hormones have dramatic effects on genital formation. Women who take hormones or other medications as part of a deliberate female-to-male transition usually experience dramatic growth of the clitoris. Use of anabolic steroids by female body-builders and female athletes can also cause marked enlargement of the clitoris together with other effects that "masculinize" the female body.

Postnatal psychosexual development is also affected by the fetal experience with testosterone. A measure of maternal testosterone during pregnancy, which correlates with fetal testosterone, is an indirect measure of fetal testosterone level. Measured in this way, a high fetal testosterone level predicts male-typical interests and behavior in adult female offspring at the age of 3.5 years. Fetal testosterone apparently controls later gender-related behavior.[6] I'll say more about this later in the chapter.

Gestation is the period during which each individual's sexuality is first expressed and shaped. But impact on the organization of fetal sexuality seems to be most effective during certain sensitive gestational periods, windows of vulnerability. The exact periods for human behavioral effects of sex hormones remain unknown. The period 8–24 weeks of gestation may be most critical because that's when testosterone secretion surges in male fetuses, but there may be multiple sensitive periods and different sensitivities in various parts of the developing brain.

Recent studies of the brains of transsexual men reveal that an important "old brain" structure involved in sexual behavior has in these men the same histological characteristics as the structure found in females.[7] The characteristics are not influenced by hormones found in the adult. It seems that gender identity, like sexuality, develops as a result of an interaction between the developing brain and sex hormones.

The result of this interaction is wonderfully various. Human sexuality (sex and gender) is essentially a spectrum that runs in a continuum from one extreme to the other extreme, from female to male. And yet, although in any individual the spectrum is malleable by both early environment and early experience from conception onward, most societies still insist that every individual be categorized as either a man or a woman—one or the other, nothing in between. Again, we run into the limits of our ability to capture in words what nature presents. We take important human characteristics—sex and gender—that exist along a continuum from female to male and arbitrarily cut the whole into two labels, "woman" and "man." When those labels don't work, or when the labels are confounded by the reality of an individual's experience, it's the individual who must navigate and somehow survive our fearful, sometimes proscriptive, reactions to that reality.

Of course, for many such individuals, survival comes at the cost of great inner turmoil, confusion about their place in society, and often devastating rejection by friends and family. Marcia is one person who is surviving just fine—now—but her journey to this point cannot have been easy, as the following stories of others who traveled a similar path can attest.

DAVID, ANNA, AND J

In medicine, a Bovie cautery device is an instrument used for electrosurgical dissection and hemostasis (stoppage of bleeding). It's like a soldering iron: an electrical current is used to heat a treatment filament or a tip; the tip becomes extremely hot and is then used to transfer heat to the tissue. The heat destroys the tissue or cuts the tissue if the tip is moved. The heat produced also stops any bleeding by burning small blood vessels. The device has limited use. It's not supposed to be used on the genitals.

In April of 1966, an 8-month-old infant boy diagnosed with a penis foreskin malformation (phimosis) that hindered urination was admitted to a Canadian hospital for circumcision. Rather than use the conventional method of circumcision, the surgeon used a Bovie device, and the infant's penis was totally destroyed by the error. The infant's name was Bruce Reimer. He later changed his name to David Reimer, and we will generally call him David. He also had a twin brother named Brian who had the same malformation of the foreskin, but following David's surgical catastrophe, Brian's surgery was cancelled and he eventually outgrew his foreskin problem.

David's parents were of course concerned about his future life and sexual happiness without a penis. They traveled with him to the Johns Hopkins Medical Center in Baltimore, where a medical group specializing in sexual development had been promoting the idea that gender identity was totally plastic during infancy and resulted primarily from social learning during early childhood. This medical team also specialized in working with children born with abnormal genitals. They thought that although they could not replace David's penis, they could surgically construct a functional vagina for him and that afterward he'd be able to achieve a functioning sexual maturation as a girl. Thus, when David was 22 months old, they surgically removed his testes and constructed labia and a partial vagina for him. He was reassigned to be raised as a female and given the name "Brenda."[8]

The Johns Hopkins medical team reported on Brenda's progress in medical journals for many years, calling it the "John/Joan' case to hide his/her identity. The medical team also continued to see the child both for hormone treatment and to assess the outcome of the case. It was considered an important test case for the idea that gender identity is a result of social learning. Its scientific importance essentially rested on two facts. First, David had an identical twin brother who shared his genes, fetal environment, and family environment; and second, this was a gender reassignment and anatomical reconstruction for a male infant with no prenatal or postnatal sexual abnormality. The medical team was optimistic about the new girl's prospects, noting in their report of the case that "maintained on estrogen therapy, she will have a normal feminine physique and a sexually attractive appearance."

When Brenda reached adolescence, she received further estrogen to induce breast development. But by age 13, when she was scheduled to receive a full surgical vagina construction, Brenda told her parents she would commit suicide if they forced her to continue visits and treatments at Johns Hopkins.

At that time Brenda was still unaware she had been born a boy. When she was 15 years old, her parents finally told her the story of her birth and sex reassignment. Apparently shocked, Brenda now refused her role as a girl. She decided to assume a male gender and began calling herself (himself) David.

By 1997, David had undergone treatments to revert to the male gender, including testosterone injections, a double mastectomy, and two operations to construct a penis. He married a woman and became stepfather to her three children. His new life did not last. On May 4, 2004, at the age of 38, David committed suicide with a sawed-off shotgun.

In interviews as an adult, David revealed that he had never been happy as a girl, that as a child he was always taunted by children at school for his tomboyish behavior, and that his schoolmates thought he was a freakish girl and called him names.

In an interview after his death, David's mother said, "I tried really, really hard to rear her as a gentle lady. But it didn't happen."[9]

Much has been written about this case, and it had considerable impact on academic theories of gender development. People who believe gender is totally determined by genetic constraints (or at least completely differentiated at birth) have used this case as evidence against the idea that gender is malleable by the postnatal environment.

The second case was first reported in 1998.[10] A genetic male-sex infant (chromosome 46 XY), during electrocautery circumcision at the age of 2 months, sustained a burn on the skin of the entire shaft of the penis, and most of the penis eventually became dead tissue and sloughed off. Five months later, the remainder of the penis and the testes were removed, and a decision was made to reassign the boy as a female and to raise the infant as a girl. We will call the girl Anna.

Two months before 11 years of age, Anna was started on feminizing hormone therapy (Premarin). When she was 12, Anna was told about the penile ablation in her infancy. At age 16, she was admitted to the hospital for vaginoplasty. At that time she denied any problems with her gender identity, and she clearly wanted to proceed with her genital repair so that she could have sexual intercourse with males. She was clinically interviewed again at 26 years when she returned to the hospital for additional vaginoplasty to relieve some discomfort during intercourse. She was in a sexual relationship with a man, but she reported having previously had three significant sexual relationships with women. She classified herself as "bisexual" and denied ever wanting to be male. She denied any uncertainty about being female from as far back as she could remember, and she reported no dysphoric feelings about being a woman. She did acknowledge certain masculine interests consistent with her "blue-collar" job, a job usually performed by men. The authors of the report describe the patient as "a tall (175 cm), thin female, casually dressed but readily perceived as a woman."

Within a few months after the final surgery, Anna and her male partner separated, and at last report she was living with a new partner, a woman, in a lesbian relationship.

The conclusions of the clinical team reporting this case are succinct: "The most plausible explanation of our patient's differentiation of a female gender identity is that sex-of-rearing as a female, beginning at around age 7 months, overrode any putative influences of a normal prenatal masculine sexual biology."

So we have two cases with different outcomes and different conclusions. Cases of this kind (technically called "ablatio penis") are extremely rare. They don't make a definitive sample, but the second case at least demonstrates that the postnatal experience can indeed overwhelm genetics and prenatal endocrine events in determining gender.

Now consider a third real case, reported in 2006, of a pair of monozygotic (identical) twins born female.[11] We call them J and L. They were born 5 weeks premature. They developed with a single placenta. J, first-born by 5 minutes, weighed at birth 4 lbs, 6 oz., and L weighed 3 lbs, 4 oz. Both twins gained weight and were healthy infants.

Behavioral differences between the twins emerged in childhood. J acted aggressively towards L. As an adult, J remembers praying to Santa Claus at the age of 3 to become a boy. Her parents also noted J's preference for male activities and clothing at that age. When the twins played house at age 3 and 4, L always took the role of wife and mother, while J took the role of husband.

During childhood, J played basketball and baseball, while L developed domestic skills. Both twins began breast development at 9 years of age. L began menstruating at age 12, a year before J. In high school, J was attracted to girls and dreamed about making love to a woman. In contrast, L was interested in boys and started dating at age 16. In despair over her gender, J attempted suicide at age 15. She began self-mutilation (cutting her breasts) at age 17, and continued this until age 25.

At age 30, J began living as a man by acquiring a male name and binding her chest. At age 32, J began transsexual hormonal treatment and at age 34 underwent chest reconstruction surgery. Meanwhile, L was living life as a woman. She married. She had her first child at age 20 and her 8th child at age 35.

At present, J is attending college and working part-time as a hotel shuttle driver. He has not yet had a phalloplasty (surgical construction of a penis) but plans to do so in the future. He says he's happier now than he has ever been. His twin sister L says she is also happy and that she has never questioned her own gender identity.

Simplistic views of human sexuality are useless. These are monozygotic twins whose sexuality evidently diverged sometime during gestation. Of course one might argue that something might have happened very early after birth, but there's no evidence of that. As we'll see later, the simplest hypothesis is that the twins' divergent sexuality resulted from differential prenatal hormone effects during fetal development. Genome identity does not mean identical gene expression, and J and L are examples of that.

What seems apparent from all these cases is that human sexuality is a good example of how genes, prenatal environment, and postnatal environment all contribute to behavior traits. As such, absolute genetic determinism of gender seems an untenable idea. Both genes and environment are involved—prenatal and postnatal environment—and how they're involved may differ in each case.

BRAIN GENDER VERSUS GENITAL SEXUALITY

What is important for us to understand is that the gender of the brain can be pushed by hormones in one direction while the sexuality of the genitals is pushed in another direction. The various interdependent hormones active during development comprise a complex system of chemical messengers—and the system is vulnerable to environmental impacts. For example, the insecticide dichloro-diphenyl-trichloroethane (DDT), in wide use since the 1960s, may be involved in the world-wide increase in the prevalence of transsexuals and homosexuals.[12] The impact of endocrine (hormone) disruptors such as DDT on fetal sexual development, gender identity, and sexual orientation is based on the influence of hormones in designing the sexuality of the embryo and fetus.

Sexual differentiation begins with the repertoire of sex chromosomes assembled at conception, but variations in sex chromosomes (XX, XY, XXY, XYY, and so on) do not directly determine gender identity or sexual orientation. Instead, the sex-chromosome role is apparently an indirect influence mediated by the secretion of various hormones during prenatal development.[13] In general, any impact on the fetal environment that causes changes in embryonic or fetal hormone profiles at any stage during development is a potential agent of change for sexual development, later gender identity, or later sexual orientation.

Transsexualism begins with the development of a gender identity at variance with the morphology of the genitals and secondary sex characteristics. People like Marcia, David, Anna, and J tell us much about the uselessness of labels like "male" and "female." These terms were useful until we began to understand a few things about the biology of sexuality and the psychology of gender identity. Now we understand that these words refer only to the extremes of a continuum of sexual and psychosocial identities. But they're still limited in what they tell us, like the words "hot" and "cold" to describe the temperature of a room: no matter what you say, the information conveyed is too vague and often not useful.

Marcia, for one, refuses to be categorized by simplistic social labels. She sees herself as a heterosexual attracted to the opposite sex because sexually she feels like a woman. She's attracted to men both physically and psychologically. Her gender identity is female, her gender role—that is, the public manifestation of her gender identity—is female, but she has no illusions that she's anatomically a woman. She will be soon, after surgery, but she's not there yet. Whatever conscious conflict ("gender dysphoria") she might have had between her genital-sex identity and her gender identity dissipated a long time ago as she realized she would never be able to define herself according to current social standards. She's determined to control her future: she will be herself—a male-to-female transsexual.

SEXUAL ORIENTATION

The question of Marcia's or anyone else's sexual orientation challenges our labels as well. The term "sexual orientation" refers to a dispositional sexual attraction towards persons of the opposite sex or of the same sex or of both sexes. That's the simple definition, but of course there's much more to it than that—including, again, the role that hormones and other factors play in determining who is attracted to whom.

Research with animals has demonstrated a major role for androgens, the male sex hormones, in determining sexual dimorphism and sexual behavior among vertebrates. This has led to the study of prenatal androgens as determinants in human sexual orientation. The current general idea is that the default human embryonic developmental path is the heterosexual female, with prenatal exposure to released androgens (primarily testosterone) diverting the embryo to the heterosexual male endpoint.

The associated idea is that homosexuality in men is due to underexposure to fetal androgens and homosexuality in women is due to overexposure of these hormones. The application of these ideas to research on the sexual orientation of adults requires "proxy markers" (for example, adult hand morphology, finger lengths, and so on, the specifics of which will be discussed shortly) for prenatal androgen exposure, and although there has been some success in correlating such markers with sexual orientation, more research is needed before the androgen hypothesis can be a secure analysis.[14]

Postnatal learning processes are also believed to be causally involved in the sexual orientation of some "femme" lesbians and some exclusive male homosexuals.[15] But there is certainly no evidence that any form of sexual orientation is simply a matter of voluntary "choice" of a lifestyle. Homosexuality, like any sexual orientation, is not a lifestyle; it's a sexuality. Some homosexuals lead a so-called homosexual lifestyle and some do not.

Nor are genes and heredity the exclusive determinants of sexual orientation, which is instead derived from a complex of variables involving genes, prenatal environment, and postnatal environment. Four real empirical situations illustrate variations of this point.

The first situation involves a pair of identical (monozygotic) twins in which one twin is homosexual and the other twin heterosexual.[16] Since the genomes of such twins are identical, such a difference in sexual orientation cannot be explained by a difference in inherited genomes. I've already discussed similar cases of transsexuality discordance in monozygotic twins.

The second situation involves a set of monozygotic male triplets, age 21 years, in which two of the triplets are heterosexual and one homosexual.[17] Again, we have identical inherited genomes with different developed sexual orientations. In both situations, other factors—pre- and postnatal environment—must somehow come into play.

The third situation involves the measurement of finger-length ratios in female monozygotic twins with one twin homosexual.[18] A "sex-dimorphic physical characteristic" is one that differs between the sexes. In humans, the 2D–4D ratio (second to fourth finger-length ratio) is a sex-dimorphic characteristic believed to reflect relative levels of prenatal first trimester testosterone. A low ratio (below 1.0) is typical of males and is associated with high prenatal levels of androgens. A high ratio (greater than 1.0) is typical of females and is associated with high prenatal estrogen levels. In female monozygotic twins discordant for sexual orientation (one twin a lesbian), the lesbian twins have a significantly lower 2D:4D ratio on both their right and left hands than their heterosexual cotwins. In female monozygotic twins concordant for sexual orientation (both with the same orientation), there are no significant differences for either hand between cotwins. These are all monozygotic twins, with each pair inheriting the same genome. It would be quite a stretch to conclude that the postnatal environment (family, culture, and so on) produces the differences in finger-length ratios. The only sensible conclusion is that variables operating during fetal development are responsible, and the likely variables for such sex dimorphism are androgens and estrogens.

The fourth situation involves a group of clinical disorders called congenital adrenal hyperplasia (CAH). These are genetic disorders characterized by inadequate synthesis of one or both of the hormones cortisol and aldosterone. In the most common disorders in this group, the consequence of this inadequate synthesis is excess androgen secretion during fetal development. Among other effects, excess androgen during development tends to masculinize the external genitalia of female fetuses. Postnatally, these girls show increased energy expenditure and "tomboyism," and at young ages they play with the toys of boys rather than with the toys of girls when compared with their relatives without the disorder.[19] Do all these girls become homosexuals? Definitely not. But their psychosexual development is definitely influenced by the excess of androgens they experienced in the womb.

Another interesting and unexpected phenomenon is the fraternal birth order effect.[20] Surveys show without equivocation that homosexual men have a greater number of older brothers than heterosexual men. The estimated odds of being homosexual increase by approximately 33 percent with each older brother.

Statistical analysis suggests that in approximately 1 in 7 homosexual men, their homosexuality is related to the fraternal birth order effect. There is no such effect in female homosexuals. Homosexual males with older brothers also have significantly lower birth weights compared to heterosexual males with older brothers. One possible reason for this phenomenon is that as mothers become progressively immunized to male-linked antigens produced by successive male fetuses, the mothers produce antibodies against these antigens. The antibodies then divert the development of the fetal brain from heterosexual male to homosexual male. But this "maternal immunity theory" remains a speculation.

Sexual orientation is one of the most complex of human behavior patterns, and we're far from any complete understanding of its origins. In any case, I doubt that in the future any set of variables will be found to be the exclusive origin of homosexuality (or for that matter, bisexuality) in every instance. There's absolutely no sensible reason why only one cause should be involved for every individual orientation.

Ultimately, the current view of both American psychiatry and the World Health Organization is the soundest public statement yet about sexual orientation: that bisexuality and homosexuality are natural sexual variations, that individual sexual orientations lie on a continuum, and that they are based on gene and/or environment-dependent neuroendocrine alterations of sexual brain organization. This set of ideas derives from hard science and represents one of the most important advances in modern science in our understanding of human behavior.

Meanwhile, what needs to be emphasized is that once again both animal and human biology point us to the conclusion that variables in the fetal environment can be important in shaping the behavior of children and adults.

EIGHT

Developmental Brain Disabilities

Approximately 4 million live births occur in the United States each year. Among those births, the approximate prevalence in the United States of various "neuropsychological developmental disabilities" in examined children—that is, those disabilities occurring in children that impair their cognitive development and function—is as follows[1]:

Attention deficit hyperactivity disorder (ADHD)	3–10 percent
Learning disability	7
Chronic emotional or behavioral problems	6
Speech defects	5
Delay in growth or development	4
Deafness	3

Blindness	1
Epilepsy	1
Autism	0.3–0.6
Cerebral palsy	0.2

An estimated 17 percent of American children less than 17 years of age are reported to have at least one neuropsychological developmental disability. That's about 1 out of every 6 children, or 11 million children under the age of 17 in America with a neuropsychological developmental disability.

Such disabilities are usually diagnosed in the school-age population, and nearly all of them endure throughout the lifetime. Their prevalence among adults is unclear, but if we assume prevalence in American adults the same as that in children (17 percent), we can estimate that more than 50 million Americans have at least one neuropsychological developmental disability.

And the prevalence of all other live-birth neonatal defects (cardiac, metabolic, and so on)? At the present time, regrettably, there is no mandatory reporting by states to a federal registry on neonatal defects, and so we must extrapolate national prevalence figures from those furnished by only some states: from 2005 data for the state of Illinois, for example, we arrive at an extrapolated estimate of approximately 5 percent of children nationwide who have been born with a defect.[2] But some reporting states have a higher prevalence while others have a lower prevalence of live-birth defects. For example, the 2004 prevalence of live-birth defects in the state of Michigan was substantially higher at 7.4

percent.[3] The prevalence of major structural or genetic birth defects in Atlanta, Georgia in 2005 was substantially lower at approximately 3 percent.[4] The equivalent prevalence in Chicago, Illinois in 2004 was even lower at 2 percent.[5]

I will focus in this chapter on major neuropsychological developmental disabilities that may involve a prenatal environmental impact: cerebral palsy, epilepsy, vision and hearing deficits, speech deficits and learning disabilities, ADHD, and, in greater detail, autism. Childhood emotional or behavioral problems will be discussed in a later chapter. Mental retardation and growth delay are discussed in the book in various places.

How many developmental brain disabilities are just the result of some biological roulette wheel, the ball bouncing around to finally land in someone's house? Which provokes the other end of the question: How many are man-made—that is, caused by some unnatural toxin that has found its way into the prenatal environment? The idea of subclinical toxicity comes into play here. As I discussed previously, the prenatal impact of environmental toxins can range from severe effects to effects barely detectable by special cognitive testing. Thirty years ago, cognitive deficits in the absence of any physical illness were reported in children with high dentine lead levels.[6] Similar results have been reported since then for children exposed in utero to polychlorinated biphenyls (PCBs) and methylmercury. But the absence of physical signs of damage in the clinic is not evidence of the absence of damage to the brain and brain function. That has to be considered in the following discussion of specific diagnosed disabilities.

CEREBRAL PALSY

Normally, visible brain white matter consists of large tracts of many thousands of myelinated nerve fibers—nerve fibers individually covered by a periodically interrupted fatty sheath that makes rapid conduction of electrical impulses over long distances possible. Rapid conduction of nerve activity to peripheral muscles is necessary for effective muscle movements, especially for movements such as walking.

A common feature of brain damage between early gestation and infancy is destruction of this cerebral white matter, the fiber tracts that connect one part of the brain to other parts or to the spinal cord. A common consequence of large-scale white matter damage is a clinical syndrome in which disturbances of motor control are prominent. Cerebral palsy is such a syndrome.

The word "palsy" is centuries old, originally denoting paralysis of some kind, but is now almost never used disconnected from the word "cerebral"—meaning the brain. Cerebral palsy refers to a group of neurological disorders that impair the ability to move or to maintain balance and posture. The neurological damage in cerebral palsy is usually a group of specific white matter brain lesions (periventricular leukomalacia) that affect the motor cortex of the brain, the control system that includes a mapping of all the voluntary muscles in the body. The symptoms produced by this damage to the motor cortex are highly variable, depending on the extent and location(s) of cortical damage. Thus, a motor disorder like cerebral palsy may be predominantly spastic (denoting involuntary and abnormal muscle contractions), dyskinetic (difficulty in performing voluntary movements), or ataxic (difficulty in coordinating voluntary movements), and may involve paralysis of all four limbs (quadriplegia) or one side (hemiplegia) or be most prominent in the legs (diplegia).

In 2000, the prevalence of cerebral palsy in metropolitan Atlanta was 3.1 per 1000 8-year-olds. Of 10-year-old children, 23 of every 10,000 had cerebral palsy. Eighty-one percent of these children had spastic cerebral palsy. Seventy-five percent had one or more other disabilities (epilepsy, mental retardation, hearing loss, or vision impairment).[7] Like many neurological and neuropsychiatric disorders, a diagnosis of "cerebral palsy" is a description of symptoms, not an identification of causes.

Indeed, the question of what causes cerebral palsy is the subject of controversy among clinicians. Is it caused by events during birth or by events during prenatal development?[8] The controversy is to some extent sustained by legal questions about medical negligence or malpractice in obstetrics. But the heart of the problem is the mistaken belief that cerebral palsy is a well-defined pathology with a single cause. Damage may certainly occur during obstetrical delivery of the newborn, but it may also occur by various means during critical stages of brain development. The general rule for fetal developing brain damage is that the same impact may cause damage in different places depending on access and critical windows of vulnerability during gestation. For similar reasons, damage in any one place may be caused by different kinds of impacts.[9]

Cerebral palsy is a salient example of fetal brain injury with many possible causes. I am not making an argument here that every case of cerebral palsy is a result of impacts on the fetal environment; only that evidence suggests some cases may indeed have such a cause. Here are some facts[10]:

1. Approximately half the cases of cerebral palsy occur in premature pregnancies, which suggests that for these cases more has happened to the fetus than a mere obstetrical delivery problem.

2. Birth asphyxia (loss of fetal oxygen during delivery) accounts for only a minority of cases of cerebral palsy.

3. Malformations of many organ systems apart from the brain are common in cerebral palsy, which suggests damage in early gestation by an environmental or physiological impact rather than an obstetrical delivery problem.

4. Familial clustering of a condition is not sufficient evidence for a genetic origin of a condition, but the absence of familial clustering suggests genetics is not involved.[11] Familial clustering is rare in cerebral palsy, which suggests inherited genetic mutations are not a significant epidemiological factor.

5. High-level fetal methylmercury exposure causes cerebral palsy, as described in Chapter 5.

6. Various maternal metabolic dysfunctions can cause fetal cerebral palsy. An example is maternal iodine deficiency and resulting hypothyroxinemia (abnormally low concentration of blood thyroxine).

7. Infections involving known and unknown pathogens may account for a substantial number of cases.

Of all the likely prenatal factors in cerebral palsy, reduced blood flow to the fetal brain (cerebral ischemia) during gestation is the most important. The second most important is apparently inflammation caused by intrauterine infection.[12] Fetal inflammatory responses can evidently contribute to brain injury by producing white matter lesions. In any case, inflammation resulting from infection of the chorio-amnion membrane system that surrounds the fetus (chorioamnionitis) is now a known causative factor for neurological disturbance and cerebral palsy.

Inflammation is a product of the immune system. As an immune response, it involves activation of immune system scavenger cells (macrophages) and secretion of inflammatory hormone-like signaling proteins (cytokines). But secretion of cytokines may also be provoked by noninfectious toxic agents—by anything that damages cells. Contact dermatitis of the skin, for example, involves an inflammatory response to certain chemicals without any initial infection.

If an inflammatory response to infection affects brain neuromotor white matter to produce lesions responsible for cerebral palsy, what do inflammatory responses to noninfectious toxins produce elsewhere in the prenatal brain? The effects of substances like methylmercury are by now well known, but less is known, if anything at all, about the effects of other toxins on the prenatal brain.

EPILEPSY

The term "epilepsy" is as clinically ambiguous as the term cerebral palsy—with epilepsy having an even older linguistic history. People who write about the history of epilepsy usually start with the ancient Greek physician Hippocrates, but its history as "the sacred disease" goes back to the ancient Babylonians and certainly before that into the mists of prehistory.[13]

Epilepsy is the general term applied to a chronic clinical presentation of various types of neurological seizures. Most people with epilepsy have experienced more than one seizure, and maybe more than one kind of neurological ("epileptic") seizure. Usually, an epileptic seizure involves a massive and synchronized discharge of many nerve cells in the brain, often in a single location. If the motor cortex is involved, its neurons massively discharge to voluntary muscles throughout the body and the typical epileptic convulsion of muscles is produced. If other areas of the brain are involved, the seizure may cause disruptions in sensation, awareness, or behavior—depending on the connections of the affected neurons.

How are the neurons affected? What we know is that they fire rapidly and together, most probably due to a radical change in local chemistry, with the chemical change often caused by a toxin or by physical trauma. Many kinds of brain damage can cause chronic seizures. We don't yet know the ultimate molecular mechanism—and various kinds of seizures may have more than one ultimate mechanism that sets off massive synchronous discharge.

Epilepsy affects about 2.7 million Americans. Approximately 10 percent of all Americans will experience a seizure sometime during their lifetime, and about 3 percent will have had a diagnosis of epilepsy by age 80.[14]

In general, any prenatal impact that causes a developmental neuropsychological disability also has the potential to be responsible for epilepsy. Cortical malformations of various types are often coupled with epilepsy in the clinic: for example, approximately 75 percent of children with magnetic resonance imaging (MRI)-recognized cortical malformations have occasional or frequent seizures.[15] But cortical malformations can be present without seizures, and there are seizure cases of epilepsy without identified cortical malformations—a typical clinical complexity when a diagnostic label is based on symptoms rather than on tangible cellular pathology. Nonetheless, we can expect that like cerebral palsy, some cases of epilepsy derive from damage to the developing fetal brain.

As with some other pathologies discussed in this book, epilepsy also shows a season-of-birth effect. An analysis of nearly 51,000 epileptic patients in Denmark who were diagnosed between 1977 and 1993 shows a deficit of epileptic births in

September and an excess in the winter months.[16] A similar result was previously obtained in a study of a large sample population in the United Kingdom.[17]

Possible prenatal impacts in epilepsy are obstetric complications, nutritional deficiencies, and particularly maternal infections during pregnancy.[18] Of prenatal infections associated with postnatal seizures and epilepsy, those due to *Cytomegalovirus*, a herpes-type virus and a common cause of human disease at all ages, are endemic throughout the world.[19] The "cytomegalo-" prefix derives from the fact that cells infected with the virus swell to a large size. Each year in America, about 40,000 live newborns are born infected with *Cytomegalovirus*, and approximately 4000 of these infants have severe congenital anomalies—including mental retardation.[20] We don't have any numbers for mild anomalies from mild infections by this virus. Moreover, most maternal *Cytomegalovirus* infections are subclinical—no symptoms in the mother. We do know, however, that in the United States, the prevalence of antibodies to *Cytomegalovirus* (a marker of exposure) is about 60 percent in the middle and upper classes and about 95 percent in lower socioeconomic classes.

An interesting situation occurs when identical (monozygotic) twins are discordant for epilepsy (one twin has epilepsy and the other twin does not) without any known cause of postnatal injury to one twin. Clinical and MRI analysis of such twins reveals that most of the epileptic twins show either lesions or anatomical anomalies not present in the normal cotwin.[21] Since the genomes of the twins are identical, something happened after the twinning of the zygote occurred, some impact that damaged the developing brain.

VISION AND HEARING DEFICITS

Since in the United States data on disease prevalence are based on irregular surveys rather than on mandatory reporting to a central agency, we don't have a full picture of the prevalence of deficits, including of those affecting vision and hearing. Prevalence does vary from one locality to another and there are disagreements about national statistics.

On the basis of available information, however, the prevalence of childhood blindness in the United States is estimated at 0.6–6.1 per 1000 children.[22] Because testing infants for visual impairments is difficult, more than 80 percent of cases of childhood visual impairment are diagnosed at ages greater than 2 years.[23] In the year 2000, the prevalence of vision impairment in 8-year-old children in metropolitan Atlanta was 1.2 per 1000. Approximately one-half to two-thirds of these children with vision impairment also have one or more other developmental disabilities.[24] According to the Centers for Disease Control (CDC), the prevalence of hearing loss in children 8 years old in metropolitan Atlanta in 2000 was the same as the prevalence of vision impairment: 1.2 per 1,000.[25] Children born with vision or hearing deficits are, in short, more common than people think.

Confusion reigns about the causes of these deficits, since some clinicians, in reporting prevalence of sensory impairments in children, classify a condition as "genetic" if there's merely a family history of that condition. But in the absence of an identified gene mutation, family history does not imply a genetic cause—poverty (and its consequences for maternal nutrition and prenatal care) is not genetic, nor are endemic maternal infections or endemic environmental toxins that impact the fetus. And in fact, infection of pregnant mothers with *Cytomegalovirus* is the leading cause of hearing loss and vision loss among children with a prenatal etiology of such deficits. Women infected for the first time during pregnancy are especially likely to transmit the virus to the fetus.[26] Our insight into other prenatal origins of blindness and deafness in children would substantially improve if we had a federal central registry for childhood sensory deficits.

SPEECH DEFICITS AND LEARNING DISABILITIES

Human speech is a complex neuropsychological performance involving perception, hearing, learning, the neuromotor system, and specific anatomical structures evolved to produce articulated sounds. Deficits can exist in the learning and use of grammar, vocabulary, speech and sound production, and in the use of language in social contexts.

The diagnostic assessment of these deficits is complex and often ambiguous, and formal diagnosis is only rarely based on quantitative measurement of a deficit. Perhaps as a result, prevalence of speech and language developmental delay is variable. For children 2–4.5 years of age, combined speech and language delay has a prevalence of 5–8 percent. Language delay alone has an apparent prevalence of 2.3–19 percent, depending on the criteria for language delay and on the population surveyed.[27]

Since speech and language acquisition develop relatively late in children, separating prenatal and postnatal effects on their development is often impossible. Known factors include the general social effects of poverty combined with fetal alcohol spectrum disorder, producing deficits in children in language performance.[28] Other factors common in a poverty environment and known to affect language performance are prenatal exposure to cocaine and tobacco.[29]

The term "learning disability" has various meanings. In the United Kingdom, for example, the term is used to describe any form of mental retardation, a below average general intellectual function, an intelligence quotient (IQ) less than 70.[30] In contrast, in the United States, the term is used to describe a specific cognitive performance level below that expected from the person's IQ—for example, a difficulty in learning how to read in a child with an average or above average IQ. My own use of the term is consistent with the U.S. definition.

One of the major learning disabilities is childhood developmental dyslexia, a difficulty with written language (for example, reading). It can occur at all levels of intellectual ability, including among children considered to be highly gifted. It is *neurologically based*, an important qualification that distinguishes it from cultural and other possible difficulties in a child's learning a language that are unrelated to any developmental neurological problems. In other words, reading difficulties alone are not a sufficient sign of dyslexia.[31]

The major problem concerning the diagnosis of dyslexia is that it does not exist as a discrete diagnostic category—as a clinical entity reliably recognized from place to place and from one time to another. Prevalence data are generally based on arbitrary cut-off points at the low end of the distribution of measured reading abilities.[32] The prevalence ranges from 17.5 percent to 36 percent of school-age children, depending on which study and which grade level is assessed.[33]

Dyslexia may be a dysfunction involving several brain systems. It's possible the deficit derives from incapacity of the brain to process brief stimuli that occur rapidly in sequence.[34] In addition, positive emission tomography brain scans have revealed a universal (cross-cultural) neurocognitive basis for dyslexia.[35] But different regions of the brain may be involved for different kinds of languages. For example, dyslexic readers of alphabetic languages such as English and dyslexic readers of nonalphabetic languages such as Chinese show important differences in which regions of the brain are active during reading. Both types of readers show thinning of the cortex in the active regions—but the regions are different.[36] This finding, coupled with the fact that the language area in the brain does not develop until after birth, refutes the notion that the brain is already hard-wired for reading language at birth. If anything is programmed by genetics in an undamaged infant, it's evidently only the capacity to wire brain circuits for language depending on inputs to the sensory systems.

In trying to identify the causes of learning disabilities like dyslexia, or for that matter of other developmental disabilities that involve postnatal behavior, many toxicologists and pediatricians focus on research results concerning the prenatal environment and its postnatal effects. The research concerning prenatal exposure in both the United States and Europe to ethanol, PCBs, and dioxins is a good example, showing significant effects on later cognitive and learning abilities during the school years.[37] The magnitude of any effects can be expected to depend on the population studied and the methods of cognitive and learning assessment.

By contrast, most behavioral geneticists tend to ignore chemical impacts in the prenatal environment and focus instead on searching for evidence of classical genetic-mutation etiologies of childhood behavioral dysfunctions. Their research on dyslexia is no exception, with twin research a popular part of their efforts, particularly in the United States, to find the genetic origins of the disorder. Using statistical (psychometric) methods involving distribution assumptions, behavioral geneticists attempt to quantify the contributions to variance of genes and environment as linear independent factors. "Genes" are considered as the set of all variables acting before birth, and "environment" as the set of all variables acting after birth. This simplistic approach ignores the prenatal environment and prenatal impacts during gestation, ignores any nonlinear interactions between genes and environment in gene expression, and even ignores important physiological and developmental differences between "identical" (monozygotic [MZ]) twins—situations in which MZ twins are discordant for dyslexia (that is, one twin later shows dyslexia, the other doesn't). Once again, such discordance is strong evidence that something important is happening after the dual inheritance of the genome produced by twinning, and it's unfortunate that the existence of such discordances are usually ignored in psychometric studies.

In short, the basic logic of such research on dyslexia has the same flaws as that of twin research on IQ[38]: It presumes that "environment" starts acting only after birth. As we might expect, the conclusions of hereditarian psychometricians who study dyslexia are about the same as those of hereditarian psychometricians who study IQ: they're hasty and unnecessary, with dyslexia dismissed as a consequence of genetic determinism, and as a more or less expected low-end-of-the-bell-curve disability. Given the neurological complexity of reading ability, and the vulnerability of that complexity during embryonic and fetal development, ignoring the prenatal environment in assessing variance in reading performance is a serious mistake.

ATTENTION DEFICIT HYPERACTIVITY DISORDER

Attention deficit hyperactivity disorder (ADHD), under one name or another, has been clinically recognized for more than a century. The diagnosis in children is based on observations of extreme hyperactivity and impulsivity with or without inattention. A better name might be hyperactivity spectrum disorder.

The diagnosis of ADHD frequently occurs when other "disruptive" behavior disorders are diagnosed. As in many other diagnoses in pediatric psychiatry, the ambiguities involved in ADHD diagnosis mean that estimates of prevalence are loose at best. The high end of prevalence estimates is 10 percent, the low end 3 percent.

There's plenty of evidence that prenatal exposure to toxins of various kinds may be a factor in postnatal ADHD:

1. Maternal dichloro-diphenyl-trichloroethane (DDT) exposure before pregnancy may be involved in some cases of offspring ADHD.[39]
2. Maternal smoking during pregnancy is a risk factor for offspring ADHD.[40] Some epidemiologists have argued that prenatal tobacco exposure accounts for 270,000 excess cases of the disorder.[41]
3. Low-level lead exposure may be an important contributor to ADHD, the effects producing less effective cognitive control.[42] According to the same epidemiologists just cited,

prenatal lead exposure accounts for 290,000 excess cases of ADHD in American children.[43] But questions do remain about whether prenatal lead exposure is more important than postnatal lead exposure during early childhood.[44] In addition, not all children with ADHD symptoms show evidence of lead in their blood, and not all children with high levels of lead in their blood have ADHD.[45]

Arriving at the ADHD diagnosis is particularly problematic when you consider that not all children diagnosed with ADHD show the same behavior, and that some toxins produce some but not all of the behavior of the ADHD disorder. For example, fetal exposure to alcohol can produce fetal alcohol spectrum disorder, but some of the symptoms of that disorder are very similar to some of the symptoms of ADHD.

In addition to toxins, the effects of postnatal childhood virus infections, meningitis, head injury, encephalitis, epilepsy, and prescription drugs have also been associated with ADHD.[46] We veer close to a diagnostic situation in which anything that damages the brain, either before or after birth, can produce the symptoms of ADHD.

The ambiguity here is in fact one reflection of our ambiguous understanding of the details of the relations in different people between what happens to the body and what happens to the mind—that is, our ambiguous understanding of the relations between phenotype cellular toxicology and phenotype behavior. But it's a miserable fallacy to assume that because we don't know the relations in detail the relations are not important.

THE ENIGMA OF AUTISM

The film *Rain Man*, first shown in 1988, brought autism into the forefront of the American psyche. But it's amazing how many prominent film critics apparently did not understand the film either intellectually or emotionally. Some of the comments of the critics seem ridiculous. There were complaints that the film had "no lasting connection to emotions" or "no particular urgent dramatic purpose." One famous critic, Pauline Kael of New Yorker magazine, called it "a piece of wet kitsch."[47] Evidently some people see no emotional content or drama in a human struggle against a debilitating mental condition.

No matter. The film won Academy Awards for Best Actor in a leading role (Dustin Hoffman as the autistic man, Raymond Babbit), Best Director, Best Picture, Best Writing—and it became the highest grossing film of the year that it appeared. The film also won the Golden Bear at the 1989 International Film Festival, and remains the only film to have won both the Golden Bear and the Academy Award for Best Picture.

The portrayal of autistic behavior in films and books is useful to increase public awareness, but such portrayals are usually of mild cases. Here are some elements of an actual case of severe autism in a Nigerian boy 13 years of age:[48]

We will call him John. He was abandoned in a refuse dump a few days after delivery and taken into a home for destitute children when his parents could not be found. As a child he rarely played with other children. He was unable to develop speech—no verbal communication. He screamed when distressed or in need of attention. He avoided eye contact, often staring into space. He never turned around when his name was called. He was disruptive with other children and often snatched their food away after finishing his own meal.

John is now 13 years old. He never utters a word. He sometimes shouts and screams without apparent reason. He hardly ever responds to instructions. He appears distant when attempts are made to interact with him. But he responds to the word "take" if you look at him while you hold a biscuit or some other snack. He'll then snatch the biscuit and eat it at once. He'll sometimes run around the room in circles, apparently with glee, and he'll stop running only when he's physically forced.

This boy has no history of any major medical illness that might have influenced his neurological development. He has a hypomelanotic skin condition (oculocutaneous albinism), but no vision or hearing impairments and no motor abnormalities. His developmental deficits are apparently restricted to communication, cognition, and social interactions.

Such is a case of severe autism.

HISTORY, DIAGNOSTIC CATEGORIES, AND PREVALENCE

Anecdotal descriptions of cases with probably some form of autism exist as early as the eighteenth century, but the term "autism" did not come into use until the middle of the twentieth century when it was introduced to describe aberrant infant behavior: lack of affective contact, a desire for sameness, fascination with objects, and mutism or noncommunicative language before 30 months of age.[49]

Today, autism terminology includes "pervasive development disorder" (PDD), or "autism spectrum disorder" (ASD), which in turn comprises five diagnostic categories: autism, Rett syndrome, Asperger syndrome, childhood disintegrative disorder, and "pervasive developmental disorder not otherwise specified" (PDD-NOS). PDD is also called "broad' autism, while the more severe autism per se is called "narrow" autism, as illustrated in John's case.

Rett syndrome is a special case that illustrates the confusion in diagnostics. It's an X-chromosome neurological disorder that is seen only in girls. It appears suddenly at about 6 months of age, leads to decreased brain and body growth with autistic behavior, severe dementia, seizures, and early death. The identical-twin concordance is 100 percent. Rett syndrome is considered part of the "pervasive developmental disorder" group, but it's probable that sooner or later it will be moved out of this diagnostic group when more is known about its etiology.

In what follows, I will usually avoid details about the prevalence and demographics of the various forms of pervasive developmental disorder because the diagnostics are unsettled enough to make the numbers uncertain. I will generally use ASD or autism to mean broad autism and use severe autism to refer to the narrow form.

At the present time, ASD is defined by significant impairments in social interaction and communication and the presence of apparently unusual behaviors and interests. Many people with ASD also have unusual ways of learning, of giving attention, or of reacting to various sensations. Their thinking and learning abilities may vary from mentally retarded to extremely intelligent.

ASD usually begins before the age of three and remains an essentially permanent condition. The condition occurs in all racial, ethnic, and socioeconomic groups, and is four times more likely to occur in boys than in girls.

As late as the 1980s, autism was thought to be a rare disorder with a prevalence of about 0.08 percent (8 per 10,000 people). In 1996, the prevalence of autism in metropolitan Atlanta was reported by the CDC as 0.34 percent.[50] As reported by the CDC, the autism prevalence rate in the United States in 2002, in a surveillance of over 400,000 children in 14 areas, was 0.66 percent.[51] The prevalence today is almost 10 times greater than the 1980s figure, and there isn't much agreement about why this is so. An epidemic? More accurate diagnosis? Greater media attention to the disorder? To date, the reasons are unclear.

What is clear is that, like many psychiatric diagnoses, ASD and severe autism are not disease entities. Rather, they are manmade labels used to denote or categorize a particular constellation of symptoms. Not all of these symptoms occur together; many of them can involve various causes; and some or all of the symptoms can be caused by qualitatively different effects on the brain. As a result, the labeling of autism forms is something of a moving target, subject to change nearly every decade in the absence of a cohesive understanding of their complex etiology.

Indeed, given the complexity of autistic behavior, it's reasonable to assume that more than a single cause might be involved in different individuals: genetic factors, prenatal impacts, postnatal toxins, or some combination of these causes. The question before us here is to what degree might the clinical entities called autism spectrum disorder (broad autism) and its severe form (narrow autism) involve prenatal impact on the developing embryo or fetus?

POSTNATAL THEORY: THE BETTELHEIM FIASCO

In the international community of psychiatrists involved with autism, the name Bettelheim rings like a dissonant cracked bell, a name many people would like to forget.

Bruno Bettelheim emigrated from Austria and arrived in the United States in 1939 at the age of 36. In 1943, he published in a psychological journal a paper entitled "Individual and Mass Behavior in Extreme Situations"—an analysis he claimed to be based on his own experiences in two German concentration camps. In the early 1940s not much was known in America about either concentration camps or people who had survived them, and with the publication of that paper Bettelheim achieved a foot in the door of American psychiatry. A year later he was appointed a full professor at the University of Chicago, where he was on the faculty for nearly 30 years and where he ran the Sonia Schankman Orthogenic School for emotionally disturbed children and adolescents.

Bettelheim believed that psychological interactions between children and parents were the sole determinants of childhood psychological development, ordinary and aberrant. He subscribed to the theory that the mothers of autistic children, for example, were "cold refrigerators" disappointed in the existence of their children. He also believed that the root cause of the emotional disabilities of autistic children was the combination of detachment and subtle hatred of the mother for her child. Science was hardly involved in these pronouncements. More important were the mantras of truth received from introspective analysis and psychoanalytic dogma. As such, to the community of psychoanalytic professionals and those in the media (many of whom were psychoanalytic patients), Bettelheim's ideas about autism reigned supreme during the 1960s and 1970s.

Bettelheim had in fact no formal training in psychiatry, medicine, or psychology. His Ph.D. from the University of Vienna was in philosophical aesthetics. He was no more qualified to diagnose or treat a brain disorder than a hawker selling sideshow tickets at a carnival. At best Bettelheim was a self-serving manipulator whose authority and career benefited from the enormous sympathy of American intellectuals for educated refugees from the Nazi regime. At worst, Bettelheim was a charlatan whose theories on autism, like something out of a Marquis de Sade novel, wrecked the lives of parents and children.[52] He committed suicide in 1990 by swallowing drugs and whiskey and tying a plastic bag over his head.[53] Shortly after his death, the Bettelheim fallacy was revealed to be a therapeutic fiasco, but its consequences are still being played out around the world.

The Bettelheim interlude in American psychiatry illustrates how dogma in science can be dangerous, in this case producing a public condemnation of the parents of autistic children, parents who suffered enormous guilt following the accusations that they were to blame for the autistic behavior manifested in their children. The best that can be said about Bettelheim is that his dogma was essentially the dogma of his contemporary psychoanalytic psychiatrists—so maybe they should all be blamed for the misery they caused.

GENE-MONGERING

These days the fashion pendulum has swung to the opposite extreme, from introspective psychodynamics to the mystique of DNA. Gene-mongering in the media is as rampant about autism as about any other aspect of human behavior, ordinary or aberrant. In 2007, in a piece about autism, ABC News quoted a pediatric neurologist:

Identical twins most of the time will both have autism. The rate of a fraternal twin having autism is zero to 10 percent, whereas for identical twins it is 80 to 90 percent. So that means that genes have a lot to do with it.[54]

Not quite. Autism in identical twins can be caused by a shared fetal environmental impact rather than by direct genetic transmission. The reality is that like mental retardation, autism is a behaviorally defined syndrome with a possibly wide variety of both genetic and nongenetic causes. The often-quoted concordance rate of 90 percent among identical twins is accurate for broad autism but not for severe (narrow) autism, for which twin concordance is only about 60 percent. Also in contrast to the ABC News report, the concordance among fraternal twins can be as high as 23 percent for broad autism.[55]

Studies of MZ (identical) twins can be important if we want a suggestion about the possible role of heredity in a condition or behavior. But the details are important. Monozygotic twins may or may not share a chorion, one of two membranes (the other is the amnion) that surround an embryo during early gestation, and the developmental consequences in each case are usually not predictable.[56] The local prenatal environment for MZ twins can be the same or different. The concordance of autism in MZ twins varies with the study and ranges from 36 percent to 90 percent.[57] Any significant concordance may imply genetic factors acting alone or genetic factors affecting susceptibility to shared fetal environmental impact. It's also possible that given a fetal environmental impact, genetic factors determine the neurological consequence—in one case autism, in another instance ADHD, and so on.

At least 60 different genetic, metabolic, and neurologic disorders have been associated with autism, including fragile X syndrome, Down syndrome, fetal valproate embryopathy, and congenital German measles (rubella).[58] But at the present time no single gene or cluster of genes has been identified as a genetic cause of a majority of cases of autism. Nor is familial autism sufficient evidence of a genetic component, since if the diagnosed familial autism is a consequence of environmental impact, familial incidence is expected. In addition, even apparent familial autism is complicated by the possibility of multiple etiologies. Given the frequency of autism in the general population, it's possible for siblings to have the syndrome for different reasons.[59]

It's also worth emphasizing (once again) that psychiatric diagnostic categories are not based on etiologies but on subjective phenomenology—descriptions of symptoms—and that whatever the genetic influences may be, complex psychiatric syndromes do not obey Mendelian genetics (that is, one discrete genotype for each discrete phenotype): different genotypes can result in the same phenotype, and the same genotype can result in multiple phenotypes.[60] Gene complexes rather than single genes are usually involved. Maybe the most important constraint is that the parameters of behavioral phenotypes are both arbitrary and continuous (traits merge into each other) and Mendelian genetics can apply only to distinctive traits.[61] In general, a Mendelian trait is a trait controlled by a single gene locus that shows a simple Mendelian inheritance pattern. Couple these facts with the multiplexing influences of environment-dependent gene expression and one caveat stands out: simple answers about the etiology of autism are unlikely. Autism is a disorder of extremely complex neurological behaviors, and so again, its causes are likely to be just as complex. The public would thus be better served by media recognition that genes, environment, and gene–environment interactions might all be involved in the etiology of autism, and that various causes might be responsible for various types of autistic behavior.

Bottom line: At the present time in the United States, the consensus is that autism is a consequence of a development diversion that occurs before birth. In about 1 percent of cases there's evidence that the diversion is already programmed by inherited genes, but there's no evidence for complete genetic determinism of the disorder in 99 percent of cases.

NEUROBIOLOGY AND EPIGENETICS

One question dominates everything: What is it that is happening in the autistic brain? Because of new instruments and a focus on the biology of the brain rather than the psychology of the mind, our knowledge of the neurobiology of autism has progressed rapidly during the past few decades. Autism is apparently a developmental neurobiological disorder involving many organ systems, including the primary autistic system—the brain. The syndrome may be a consequence of dysfunction of the brain region called "association cortex" (various regions of cortex that apparently integrate activity of primary sensory or motor functions)—both of neurons in association cortex and of projections into and out of it.[62]

MRI studies of adolescent autistic children show increases in thickness of the cerebral cortex (gray matter) in specific areas of the brain coupled with decreases in thickness of fiber tracts (white matter) leading in and out of those and other areas.[63] This provokes the idea that a combination of developmentally enlarged cortex and reduced white matter may be the structural basis of autism. But it's also possible that the anatomical differences may be a consequence of autistic behavior rather than a basis, since the development of the brain continues throughout childhood.

Can autistic behavior be related to specifics about nerve cells? A neuron "fires" nerve impulses either spontaneously or as a result of input of nerve impulses from other neurons. "Mirror" neurons are those in the brain cortex of animals (and presumably also in humans) that fire both when movements of the body occur in the animal and when the animal observes movements in another animal.[64] Several regions of the animal cortex apparently contain mirror neurons, and in the analogous regions in human brain cortex of autistic patients, neuron activity is evidently depressed. This has led to the idea that autism is caused by a lack of mirror neurons, the deficiency leading to deficits in social skills, imitation, empathy, and so on. The idea is speculative. Other mechanisms that would produce autistic behavior are possible. At the present time we have no experimental data that definitively identify mirror neurons and their activity in the human brain.

A different neurological view of autism involves brain region disconnection. The autism developmental disconnection hypothesis holds that association areas of the brain involved in higher functions that normally connect to the frontal lobes are partially disconnected during development in autism.[65]

Some researchers also suggest that there might be at least two different developmental pathways to autism, one related to primary connections between the temporal lobes and frontal lobes (late prenatal origins), and another pathway related to brainstem dysfunction (early prenatal origins).[66]

Another idea is that prenatal impact on the activity of the neurotransmitter dopamine, including the effects of maternal psychosocial stress, maternal fever, maternal genetics and hormonal status, use of certain medications, and fetal hypoxia, may be involved in the epigenetic etiology of autism.[67]

The importance of epigenetics in the etiology of autism is suggested by the occurrence of autism in patients with autistic disorders that apparently arise from discrete epigenetic mutations (for example, fragile X syndrome), or that involve known epigenetic regulatory factors (for example, Rett syndrome).[68] Impact on epigenetic regulation can produce dysfunctional switching of genes on or off, including genes responsible for detailed nerve cell organization in the brain.[69]

OTHER CAUSAL POSSIBILITIES

Advanced maternal and paternal ages produce an increased risk of autism in offspring,[70] but they have also been related to miscarriage during pregnancy, childhood cancer, autoimmune disorders, and schizophrenia. The parental age effect is not, in other words, specific for autism.

An interesting correlation exists between ASD in children and psychiatric disorders in their parents,[71] who are, for example, more likely to have been hospitalized for a mental disorder than parents of control subjects. Schizophrenia is more common among parents of autistic children than among parents of controls. Depression and personality disorders are more common among mothers of autistic children but not more common among their fathers.

Reanalysis of some older data has revealed that a significant relation does exist between blood levels of mercury and a diagnosis of autism. Hair sample analysis of mercury suggests that people with autism may be less efficient and more variable in eliminating mercury from blood.[72]

The vaccine-mercury linkage, recently under heavy suspicion as a cause of postnatal autism, is no longer considered significant. The reason given is that the cessation of the use of postnatal vaccines containing mercury (in the compound thimerosal) as a preservative has produced no decrease in the prevalence of autism.[73] But there might be other reasons for the absence of a decrease in prevalence other than an absence of linkage.

The vaccine issue is not closed, but it may not be possible to finally resolve it because different causes of autism may be at work in different cases.

PCBs are known to be potent immunotoxins, producing in many animal species atrophy in the thymus gland, a major gland of the immune system. Autism is characterized by immune system parameter deviations, and so it's possible that autism may be a consequence of an autoimmune process in the developing brain resulting from prenatal exposure to PCBs.[74]

A positive association exists between mothers who live near organochlorine pesticide use and the incidence of autism in offspring.[75] The incidence declines with increasing distance from agricultural fields that use such pesticides.

Maternal iodine deficiency might also be associated with fetal vulnerability to pesticides.[76] According to this idea, pregnant women who have marginal or deficient iodine nutrition due to exposure to pesticides may induce iodine deficiency in the fetus and consequent negative effects on the developing fetal brain. The iodine nutrition status of Americans has been declining over the past three decades and is currently borderline deficient in 30–40 percent of pregnant women.[77] Mothers of autistic children tend to be non-Hispanic white and non-Hispanic black women, and these two groups have the poorest iodine nutrition status in the United States.

The autism syndrome can apparently be caused by a brain infection. For example, the viral brain disease herpes encephalopathy can produce all of the main symptoms of autism, although sometimes the symptoms are reversible when the infection is cleared.[78]

In addition to acute infections, chronic infections, such as the tick-borne Lyme disease (caused by the microbe *Borrelia burgdorferi* and transmitted by the deer tick), can have direct effects on or promote other infections in the developing fetus by causing fetal immunological vulnerability. As I've noted, the chronic immune response to infection is a cluster of biochemical changes that can affect development of the fetal brain. Many pregnant women with Lyme disease have offspring with ASD.[79] And the symptoms of tick-born diseases like Lyme disease are often similar to the symptoms of ASD. Approximately 20–30 percent of ASD disorder patients apparently are or have been infected with the Lyme disease microbe, and the symptoms of many children with ASD often improve with antibiotic treatment. If not every case of autistic spectrum disorder involves chronic fetal infection, there is certainly enough evidence to warrant close inspection of possibilities.

What is apparent from all of the facts and complexities is that tabulating the possible causes of autism only increases the puzzle. But if we remember that any impact that affects the development or physiology of the fetal brain has at least the potential to cause autistic symptoms to be manifest after birth, maybe the puzzle is less dramatic.

A CAVEAT ABOUT CORRELATIONS
AND CAUSALITY

The collection of possible causes, consequences, prevalences, and correlations of various brain disabilities in this chapter reminds us that we need to be cautious about labels and numbers. Of particular importance is the necessity for caution about correlations, and a caveat is appropriate here, a caveat also appropriate for other parts of this book.

A correlation between two variables is not evidence of any causal relationship between the variables. Although this is one of the fundamental maxims of scientific data analysis, it's usually ignored by the media and too frequently also ignored by researchers themselves.

It's not unusual to be confronted with a specious argument that correlations can imply causal relations in behavior genetics because genetic variations are "natural experiments" of evolution and heredity and environmental variations are "experiments of society"—so that correlations between one and the other do suggest causes and effects.[80] Such an argument seems a twisting of language to avoid the mathematical realities of statistics. A correlation can be said to reflect causality only when we already deduce from experiments that causality exists. It does not work backwards: given a correlation, we can say nothing about causal relations between correlated variables—no matter whether we imagine our variables to be "natural" experiments. For example, suppose we're presented with a high correlation coefficient between two variables, A and B, and the variables are not identified. There is no way we can say that it's A that causes B or B that causes A or neither of these possibilities and merely correlations with a third unknown variable that may be causing both A and B. But suppose we then learn that A is the incidence of malaria in a number of locations and B is the population of mosquitoes in those locations. Now things are different: we know mosquitoes are carriers of the malaria parasite, so that given the correlation we can say the correlation does reflect causality.

But with that said, we must also understand that the maxim is not a reason to ignore the correlation. A correlation is a hint that something important may be connecting the two variables, and in the domain of toxicology and public health, a hint should be enough to warrant a closer look.

The existence of other studies that fail to show the correlation should not deter the closer look. In public health too much is at stake. This is the general attitude of the Centers for Disease Control and Prevention and it's my attitude in this book. It means I am more interested in evidence of effects than in no evidence, since the high stakes make the mere existence of any evidence important.

Differences in geography or in population sampling are often the explanations for correlation differences. Consider the Japanese Minamata methylmercury catastrophe as an example. The fact that there has never been an episode like that in the United States is not a counterargument against the idea that

prenatal methylmercury exposure can produce serious dysmorphologies in the fetus. In general, in the domain of toxicology
and public health, whatever evidence exists is more important
than reports of no evidence.

NO SIMPLE ANSWERS,
NO QUICK FIXES

Before 1950, autism hardly existed in public awareness. After
about 1950, there was much media talk about autism, speculations about its origins, and much unfortunate ballyhoo by Bruno
Bettelheim and his followers, with blame heaped on parents
who were thought to be the cause of autistic children by with-
holding love and nurturing. As I have indicated, the accusations
were a tragic error that often had tragic (and expected) consequences resulting from heaping blame on parents, especially on
mothers. Many of us who lived through the Bettelheim years
know of cases of misery, divorce, and even suicide provoked
by the accusations of parental responsibility. Now there's been
some progress. In recent years, biologically oriented psychiatry,
plus film actors like Dustin Hoffman (*Rain Man*) and Sigourney
Weaver (*Snow Cake*), actors with prodigious talents, have made
the public aware of the biological and psychological subtleties
and complexities of autism.

Still, too many people do not understand that autism is not
a well-defined neurological problem with a single cause. Always

to be emphasized in any litany of possible etiological factors, such as those identified in this chapter, is that what we call "autism" can have different causes in different people. There are many ways to disrupt the brain to produce lasting autistic behavior, and many more ways to cause disruptions during prenatal development of the brain and nervous system. The possible prenatal causes of autism are as subtle and as complex as autistic behavior itself. There are no simple answers and quick fixes. And any attempt to find them will likely be a search down a blind alley.

The current consensus in the American medical community is that autism spectrum disorders involve a complex relationship between exposure to environmental stressors and genetic susceptibilities.[81] That doesn't tell us much, but it's certainly an advance over the theory that the mothers of autistic children were emotional refrigerators.

Will we ever unravel the enigma of autism? Most certainly, yes. As long as science is free to move forward, the puzzle of autism will eventually be solved.

NINE

GENES, THE WOMB, AND MENTAL ILLNESS

The term "mental illness" is about as ambiguous as you can get, a label that covers everything from mild mood problems to complete derangement in violent paranoid schizophrenia. My focus in this chapter is on those mental disorders for which there is some evidence relating the condition to impacts on the fetal environment. That leaves out many psychiatric conditions known to clinicians, and even many common kinds of mental illness. So the absence of any discussion of some specific mental disorder in this chapter does not mean it's unrelated to the fetal environment—it means only that I know of no sufficient evidence to present here. My purpose here is to show that there's enough scientific and clinical data to suggest there may be important connections between the fetal environment and later mental disorders. The major sections in this chapter discuss psychosis and nonpsychotic disorders, with schizophrenia the main subtopic in psychosis, and major depression and eating disorders the main subtopics in the second section.

A word about prevalence: mental illness is more common than many people imagine. The current prevalence estimates are that about half the U.S. population meets the criteria for at least one mental disorder during a lifetime, with about 25 percent of the population meeting the criteria for at least one mental disorder during any given year.[1] Of these disorders, the most prevalent are apparently anxiety disorders, followed by mood disorders (for example, major depressive disorder), impulse-control disorders (for example, attention deficit hyperactivity disorder [ADHD]), and substance disorders (for example, alcohol abuse). In contrast, the prevalence of psychosis as I define it here is only 2–3 percent of the U.S. population, and the world prevalence is about the same.

In this chapter, as in the previous chapter, we have problems with labels. Diagnostic terminology in clinical practice often complicates our understanding of the scientific literature on mental disorder. Many psychiatric diagnoses—labels that describe which symptoms characterize a psychiatric disorder and distinguish it from another—have varied with time and place over many years, with much debate still going on to this day about where to draw the line between the normal and the abnormal and between various states of the abnormal. Ordinary behavior in one culture may not be considered ordinary behavior in another culture. A psychiatric diagnosis in America may not match a psychiatric diagnosis of the same person in Jordan.[2] A more specific example concerns the current diagnostic dichotomy that separates schizophrenia as a discretely different illness from bipolar disorder, a labeling antique derived from nineteenth century psychiatry. Today, this dichotomy is wearing thin as many psychiatrists realize that with the advent of molecular neuropsychiatry the labels for these mental disorders and others are losing relevance.[3] I will try to adhere to current psychiatric diagnoses in this chapter, but I will nonetheless point out the difficulties they pose. In general, the ambiguities and overlap between the symptom clusters used for the diagnosis of mental illness are frustrating to anyone interested in scientific precision.

PSYCHOSIS

In plain language, psychosis is madness. Another term for it is insanity—what the nineteenth century psychiatrist Heinrich Neumann believed arose from a "loosening of the togetherness"— the mind coming apart.[4] This phrase sounds ludicrous, but as an intuitive idea of the nature of madness, it's probably on the mark.

The trouble with understanding madness is that for every way to unhinge the brain the result may be ten ways to unhinge the mind—with consequences that can range from great art to great social chaos. Between these extremes, though, are the more typical consequences that most people with severe mental illness experience: not extraordinary creative expression that enriches the world, not unspeakable acts of violence that turn the world on its head, just the great personal pain brought on by one's private demons and by one's inability to cope with the world and a life gone awry.

I use the word "psychosis" in this section, but in fact it's a word that continues to be used despite increasing ambiguity about what it means. Technically, the term psychosis currently refers to any mental disorder characterized by delusions and/ or hallucinations, especially when there's an apparent inability to distinguish between reality and fantasy. Schizophrenia, for example, is a psychosis under current terminology. An older definition of psychosis is any mental disorder with an apparent inability to distinguish between reality and fantasy with or without delusions or hallucinations. In the early days of modern psychiatry, a great deal of energy was devoted to classifying mental disorders, but the recent trend has been to recognize that boundaries between diagnostic categories are often diffuse, and diagnostic categories themselves can cause a great deal of public and professional confusion. For example, a person with an eating disorder like anorexia nervosa may literally starve himself or herself to death for a fear of gaining weight even when they are emaciated to the endpoint of life. In psychiatry, this is an eating disorder, not a psychosis, even if the condition involves bizarre self-destruction based on body-image fantasy. Psychiatry does have its label problems. The famous quip by the American psychiatrist Thomas Szasz is worth repeating[5]: "If you talk to God, you are praying; if God talks to you, you have schizophrenia."

This section is mostly about schizophrenia because there's intriguing evidence relating the fetal environment to schizophrenia, and not much research yet for most other psychoses.

If psychosis is madness, then schizophrenia is the Queen of Madness, complex and always terrifying. With the flowering of medical psychiatry in the late nineteenth century, the character of the disorder was described and named, and the modern description is not much different than the description popular 100 years ago: behavior that includes delusions, hallucinations, disorganized thinking and speech—bizarre and inappropriate behavior. Its age of onset is usually in late adolescence or early adulthood, although precursor ("prodromal") signs of the disorder may begin to manifest themselves some years before the first psychotic episode. After that, if untreated, the illness in most cases produces a progressive deterioration in the individual's ability to function. Among its worst effects is that it almost always alienates the individual from others: As one schizophrenic patient once told me, his condition was like a dark solitary confinement without walls. Things that made sense to him apparently made no sense to other people. He was always alone. That he is among 3 million other Americans who will be diagnosed with schizophrenia during their lives (and among 60 million worldwide)—an estimate representing a lifetime prevalence (probability of the diagnosis at any time during a lifespan) of approximately 1 percent of the population[6]—would be cold comfort to him.

Psychosis is a feature of other mental disorders as well. One is bipolar disorder, a serious mood ("affective") disorder, also known as manic-depressive illness, that includes several subtypes depending on the occurrence and the degree of severity of their mood symptoms. An individual with the most severe subtype of bipolar disorder (bipolar I), for example, can cycle from extreme depression to extreme euphoria in a single day, the depression sometimes producing an urge to kill oneself, the euphoria some-times producing megalomania, among other symptoms; at both extremes, psychosis in the form of delusions and hallucinations can occur. The typical age of onset for bipolar disorder, as for schizophrenia, is in late adolescence or early adulthood.[7] Its life-time prevalence in the United States is approximately 2 percent. Its ordinary prevalence in American adults is approximately 2.6 percent—which corresponds to approximately 6 million people over the age of 18.[8]

Psychosis can also occur in what's known as schizoaffective disorder, essentially a symptom-combination of schizophre-nia and bipolar disorder seen in some patients. In a recently conceived diagnostic categorization, schizoaffective disorder is included with schizophrenia, along with related disorders such as schizotypal and schizoid personality disorders, under the umbrella term "schizophrenia spectrum disorders." The lifetime prevalence of schizoaffective disorder is less than 1 percent, but the numbers are flexible, since it's often used as a preliminary diagnosis when there's uncertainty about which label to use for a patient.[9] The depressive form of schizoaffective disorder is more common in older persons, while the bipolar type is more common in young adults.

It's unfortunate that despite the significant prevalence of bipolar disorder and schizoaffective disorder, we have only scattered data (some of it discussed later) about their connection to fetal impacts. The general research problem relating fetal environment to adult psychosis is the often long delay between gestation and onset of mental illness. For example, research comparing adult mental illness in the cohort gestating around the World Trade Center during the 9/11 catastrophe to the general population will probably not begin until 2026—25 years after the event.

Possible Causes of Schizophrenia

Nineteenth century pre-Freudian psychiatry proposed an "organic" (neurological) cause for schizophrenia. When Freudian psychology became popular in the twentieth century, the focus of clinicians treating schizophrenic patients shifted to psychogenic causes, with nearly everyone in the psychiatric community speaking in one voice against the idea of a neurogenic etiology:

> The organic nature of schizophrenia has not been demonstrated. Genetic studies have failed to reveal, in a clear and uncontroversial manner, that schizophrenia follows certain genetic laws. Anatomical pathology, including neuropathology,

has not ascertained any causal relation between organic factors and schizophrenia.[10]

So they turned away from biology to psychodynamic causes, to family environment, particularly to early childhood confusion between reality and family dogma. That lasted until psychiatry rediscovered biology in the 1960s. The rediscovery depended not on the new interest in genes and DNA but first on the practical evidence that various new drugs seemed useful in treating schizophrenia, coupled with expanding data about how these drugs affect the interactions of nerve cells with each other. Neurochemistry became the prime focus of biological psychiatry, and soon attention shifted to genes that might control the neurochemistry of the brain.

Advances in neurochemistry and genetic analysis have forced revisions in thinking. For example, the evidence from modern neuropsychiatry, especially from genetic neuropsychiatry, does not conform to the classical idea that schizophrenia and bipolar disorder are two separate disease entities. Research is now revealing considerable overlap in both the genes and gene expression of these conditions and it's likely that as a result the next overhaul of psychiatric diagnostic categories will see important changes in the way psychoses are diagnostically categorized.[11]

But the now increasing overemphasis on genetic determinism in psychosis has led to labeling various mental disorders as "genetic diseases" when there is no evidence that their cause is merely the inheritance of a genetic mutation. This is certainly true for schizophrenia. Many biologists are unhappy with such simplistic gene-mongering.[12] So-called genetic psychiatric disorders usually have many identified genetic determinants. In addition, their clinical outcome always depends on environmental circumstances that begin with conception.[13] Thus the rationale for the emphasis on chromosome "markers" is apparently flawed.[14] Chromosome markers for psychiatric disorders, including schizophrenia, routinely turn up on different chromosomes in different cases.

Neurodevelopmental models of schizophrenia have for the most part emphasized prenatal brain development, focusing on effects that may alter gene expression.[15] For example, according to one idea, individuals with schizophrenia inherit genes that cause structural brain deviations that may be compounded by early environmental impact.[16] Much of the focus of this hypothesis has been on postnatal impact, such as stress, that changes brain chemistry, but the idea can also accommodate a focus on prenatal impact.

Another idea is that the genes involved in schizophrenia influence synaptic plasticity and the development and stabilization of neuronal microcircuitry in the brain. Influences on synaptic plasticity would involve specific synaptic transmitters, such as dopamine and gamma-amino butyric acid (GABA).[17]

Studies of twins demonstrate that schizophrenia is a complex syndrome involving both genetic and environmental influences. Some researchers like to use the term "trait" rather than syndrome. But given our still ill-defined views of psychosis, it's probably prudent to avoid any such term that suggests that schizophrenia is a succinct clinical entity. The genetics of schizophrenia is certainly not Mendelian genetics wherein discrete traits are always linked to discrete genes and the links always obey the various Mendelian laws of heredity.

Twin research in schizophrenia presents us with puzzles. For example, although the usual quoted figure for concordance of schizophrenia in monozygotic (MZ) twins is approximately 50 percent, the classical modern study on the subject shows a variation of concordance of 31–58 percent depending on which Western country provides the statistics. The concordance rate in the United States, for example, is only 31 percent, while in the United Kingdom it's 58 percent.[18] That is not a minor difference.

So the discordance for schizophrenia between some pairs of MZ twins (that is, one twin has schizophrenia, the other doesn't) can be as high as 69 percent (the figure for the United States). I think this is a more important fact than any estimates of "heritability" (usually about 80 percent for schizophrenia) based on the customary questionable assumptions in the behavioral genetics of twin research. One such assumption is that there is complete linearity of the heritability equation—ignoring, especially, gene–environment interactions during gestation.[19] In general, the so-called "heritability" of schizophrenia is as useless a calculation as the heritability of intelligence quotient (IQ) (see Chapter 10). It's based on problematic statistical assumptions about correlations and the origins of variance. Current calculated heritability is a function of the populations studied, the postnatal environment, the phenotypic criteria used, the statistical model used, and many other factors.[20] The heritability measure does not get us very far in understanding schizophrenia.

In contrast, the high discordances in some MZ twins, particularly for schizophrenia, are telling us something about origins—that something is happening during fetal development that can trump genome identity.[21] This suggests that avoiding general consideration of the fetal environment in estimating the contribution of heredity to schizophrenia is a serious mistake; the discordance evidence is telling us that the fetal environment is important.

The best view we have from twin studies is that schizophrenia results from both genetic and shared-environment etiological influences, with "shared environment" meaning classical postnatal environmental factors plus prenatal environmental factors, such as exposure to infectious agents, macro- or micronutrient dietary characteristics, and exposure to environmental toxins, teratogens, and other assaults to the prenatal environment.[22] Which factors and assaults are important can differ from one individual to another for reasons still unknown. Schizophrenia is not a simple puzzle.

If discordances in MZ twins for psychosis imply that environmental differences are involved, the discordances may be caused by epigenetic differences. One idea is that epigenetic differences in MZ twins are stochastic, due to random effects altering gene expression.[23] Another idea is that differing intra-uterine environments in MZ twins can lead to discordant consequences.[24] The effects of differences in placentas of MZ twins illustrate how uterine environment can affect development in schizophrenia.[25]

Psychosis and Season of Birth

If the prevalence of an adult mental illness is correlated with season of birth, the simplest explanation is that something happened during gestation. We know of no biological mechanism for season of birth to change an inherited genome, but we know plenty of possible mechanisms for season of birth to affect fetal development by infection, nutrition, stress, and so on.

Many studies are consistent in showing a 5–8 percent winter-spring excess of births for both schizophrenia and bipolar disorder. A birth excess is also found in schizoaffective disorder (December–March), major depression (March–May), and autism (March), but not in other psychiatric conditions.[26] Parental procreation habits have been discounted. The explanations that have been offered are seasonal effects of gene expression, subtle pregnancy and birth complications, ambient light, internal chemistry, toxins, nutrition, temperature or weather, infectious agents, or some combination of any of these causes—the lengthy repertoire of possibilities essentially defining our ignorance of what happens. Our inability to explain such seasonal correlations underlines how little we know about the etiology of mental disorders.

An analysis of the Dutch Hunger Winter of 1944–1945, described in chapter 5, reveals that the risk of schizophrenia peaked in the cohort that was in gestation during this period. This suggests that nutritional deficiency of pregnant women may play a role in the origin of later schizophrenia in their offspring. Similar results are obtained for affective disorders in the same cohort.[27] In many populations, maternal nutrition during pregnancy changes with the seasons. Epidemiological studies show a correlation in populations between the incidence of schizophrenia and the incidence of neural tube defects, which also suggests that maternal nutritional deficiency, a known cause of fetal neural tube defects, may be involved in offspring schizophrenia.[28] But variations in maternal nutrition rather than nutritional deficiencies may be involved in the link to season of birth.[29] As for geography, the "winter-born" correlation for schizophrenia is also found in the southern hemisphere, where the "winter months" are July, August, and September.

The season-of-birth link to psychosis remains a puzzle. One possibility is that winter-related viruses may be involved.[30] Viral infections during pregnancy, discussed in the next section, can certainly impact the development of the fetal brain.

Fetal Infection, Fetal Hypoxia, and Other Impacts

A study of 20,000 pregnant women in California, begun in the 1990s, has been yielding data about the relation between prenatal environment and the risk for adult onset schizophrenia spectrum disorders (SSDs). Results indicate that first trimester fetal exposure to the influenza virus produces a 7-fold increase in the risk of later schizophrenia and SSD.[31] This is a strong confirmation of earlier results relating maternal influenza infection to offspring schizophrenia.[32] According to one hypothesis, this outcome may be due to severe maternal hyperthermia, a symptom of influenza infection and a known cause of congenital defects of the central nervous system of the fetus.[33]

In general, prenatal exposure to infection apparently increases the risk of later schizophrenia and other neurodevelopmental disorders. As discussed previously, the release of proinflammatory cytokines during maternal infection may have a damaging impact on fetal brain development.[34]

Tabulating the occurrence and degree of fetal hypoxia in past births is usually done by examining hospital records and by interviews—indirect assessment at best. But most studies agree that MRI scans of the brains of adult schizophrenics show greater brain abnormalities for those patients who apparently experienced fetal hypoxia during gestation, and that fetal hypoxia predicts early onset schizophrenia.[35]

The expression of many of the genes identified to be related in some way to the appearance of adult schizophrenia is known to be changed by hypoxia. This suggests that hypoxia during fetal brain development may be somehow involved in the etiology of schizophrenia.[36]

Maternal psychological stress during pregnancy most certainly results in transient or enduring physiological and hormonal changes that may affect fetal brain development. So the theoretical principle concerning maternal stress is clear. In practice, however, maternal stress is extremely difficult to measure, as is its connection to the later onset of a disorder that, like schizophrenia, may not be diagnosed in offspring until 20 years later. That's the general problem of research attempting to relate prenatal impacts to adult pathology.

But despite the research limitations, a relationship between maternal stress during pregnancy and later adult schizophrenia in offspring has been established at least in outline.[37] Considerable evidence shows that maternal stress during the first trimester of pregnancy may be a particularly important risk factor for later onset of this disorder in offspring.[38]

Other possible prenatal factors are neurotoxins: Although there has been too little research examining schizophrenia as a possible outcome of fetal lead exposure, it has been proposed that fetal blood levels of lead greater than 15 units (micrograms per deciliter) may double the risk of childhood or adult schizophrenia spectrum disorder.[39] Another important possibility is that prenatal exposure to endocrine disruptors, such as bis-phenol-A (a polymer environmental pollutant present in many plastic consumer products) can be involved in the etiology of schizophrenia.[40] Past research has indeed demonstrated endocrine and neuroendocrine abnormalities in schizophrenics.[41]

A postnatal oddity is worth mentioning here, an oddity that illustrates the complexity of the etiology of a psychosis. It's the puzzling relationship between celiac disease and schizophrenia. Celiac disease is an inherited disease involving malabsorption in the small intestine and an inflammatory intestinal response to gluten. Individuals with a history of childhood celiac disease have an increased risk of adult schizophrenia, for reasons unknown but maybe involving the inability of the celiac patient's intestine to block neurotoxins.[42]

Socioeconomic Status

It's common almost everywhere to find an inverse relation between psychiatric disorders and socioeconomic status. Does the stress of low socioeconomic status provoke psychiatric dysfunction or does the presence of psychiatric dysfunction drive the individual to a lower status? The question is generally without an answer, although many studies suggest that the stress produced by low socioeconomic status has an impact on already susceptible individuals.[43]

Certainly, low socioeconomic status increases the prevalence of infection, exposure to toxins, poor nutrition, and many other prenatal impacts that may be involved in fetal brain damage and in the etiology of later mental illness.

A feedback mechanism may also be at work: low socioeconomic status lowers the quality of prenatal care, which increases the probability that results of poor care (for example, nutritional deficiencies) will affect gene expression. This increases the liability for later psychiatric disorders, whose deficits in turn maintain the individual in a low socioeconomic status—a transgenerational nongenetic inheritance of psychiatric disorder.

As always, it's not theoretically necessary that the same mechanism be involved in a psychiatric disorder in every individual or group.

We'll return in depth to socioeconomic status and its impact on fetal development in chapter 11.

FETAL ENVIRONMENT, TEMPERAMENT, AND NONPSYCHOTIC MENTAL ILLNESS

As I've indicated in several places in this book, we're limited by our data, by what's practical and possible in research. This section is about the nonpsychotic mental illnesses major depression and eating disorders, illnesses for which there is enough evidence relating fetal environment to adult pathology. I have also included some discussion of research on temperament, because temperament is a marker of mood, and there happens to be data relating fetal environment to the temperament of infants a short time after birth. Infant temperament is a major part of our knowledge of infant behavior, since infants are for the most part not yet ready to be subjects of traditional psychiatric examination. I'm including research relating fetal environment to infant temperament here because it tells us something about how the fetal environment may shape later emotional behavior—a shaping that has to be of sharp relevance for later mood disorders.

Anxiety disorders are a major clinical entity, but there's hardly any unambiguous information yet relating such disorders to the fetal environment, so there's nothing to discuss here.

To a physiologist, the neuroendocrine system is expected to be of importance in both psychosis and especially in nonpsychotic mental disorders. We already have evidence that physiological changes in fetal neuroendocrine systems may predispose people to later cardiovascular disease by an influence on risk factors such as plasma glucose, plasma lipid concentrations, and blood pressure. Is it possible that fetal programming of physiological stress susceptibility can combine with childhood and adult psychological stress to produce psychological dysfunction, such as in mood disorders? The answer is yes. Prenatal exposure to excess glucocorticoids (for example, cortisol, the stress hormone) can program the autonomic and central nervous system so that the adult is predisposed to psychological dysfunction.[44] Meanwhile, later stress-related mood disorders may involve insufficient glucocorticoid signaling.[45] Since postnatal glucocorticoid levels have been shown to be reduced by fetal programming, a possible theoretical connection now exists between fetal programming and mood disorders. But the linkage between maternal cortisol and fetal and postnatal cortisol is not always predictable.[46]

The major mechanism that may be operating to connect maternal stress during pregnancy to postnatal psychopathology involves effects of maternal stress hormones on development of the fetal brain, particularly of the fetal hypothalamus–pituitary–

adrenal axis.[47] But what we know about human fetal development is, at the present time, limited by our technology. One significant marker is birth weight, which can be measured precisely. A retardation of fetal growth is a known result of maternal stress during pregnancy, which in turn results in different postnatal effects. Low birth weight in full-term, apparently healthy, newborns is correlated with later development of certain temperaments. Small body size in an infant full-term at birth is a predictor of adult depression.[48] Low birth weight in preterm ("premature") infants is correlated with the development of postnatal emotional problems, and physical pathology, in children.[49] So we have some signposts directing us in our search for understanding.

Temperament and Mood

Definitions of "temperament" vary in psychology, but generally, it's an overt, observable, and more or less constant behavioral predisposition. In classical psychology, temperament was the innate aspect of an individual's personality, that part of the personality that was genetically based. This definition is no longer useful, since researchers now realize that any temperament recognized at birth or infancy can be a result of impact on the fetus during development and thus need not be inherited.

Ultimately, temperament is just a label of behavior, and like all such labeling, the outcome is often subjective and dependent on who is doing the labeling and on what psychological scales—which differ from one generation to the next and from one culture to another—are used to assess temperament.

Negative temperaments characterized primarily by emotions of fear, discomfort, anger, or sadness are the most easily recognized in infants and children. What is actually observed through assessments of these temperaments are the child's reactions to stimuli—new or old. The research has its limitations (for example, qualitative judgments, possible observer bias), but the questions that are asked are important and the results can be useful.

For example, some infant or early childhood temperaments, such as an inhibited or negative emotionality, have been found to be possible precursors of psychopathology in later childhood. Children with dysfunctional temperaments may also provoke behavior from their parents that consolidates the dysfunctional temperament.[50]

Early postnatal temperament differences are already present during the prenatal period as evidenced by fetal heart rate and motor activity during experiments involving deliberate maternal emotional activation by the researcher.[51] Physiological markers of individual differences in infant temperament can be identified during the fetal period, and these markers may be shaped by the fetal environment.[52] Lower levels of fetal hemoglobin and serum iron are related to higher levels of negative emotionality (affect) and lower levels of alertness and soothability (the ease with which an infant can be calmed down).[53]

Does early childhood temperament predict adult temperament? The evidence is that behavioral styles at age 3 years are linked to personality traits at age 18 years.[54] This won't surprise most parents, who themselves will note that the personality traits of their children as young adults are often already in place at an early age.

The physiology of temperament is connected to the physiology of emotions. The part of the human brain of most importance in emotions is the limbic system; this is true also for mood, a term that in psychology refers to a more or less sustained emotional state. The limbic system of circuits operates by sending signals outside the brain to the endocrine and autonomic nervous systems, and itself receives inputs from other brain areas, including those involved in pleasure, pain, motor function, executive function, and so on. The neurological substrates of emotions and mood are highly sensitive to neurotransmitter changes, especially changes in the distribution and local concentrations of the neurotransmittter dopamine.

But while we know in general which parts of the brain are involved in mood, we know very little about how the parts are involved and next to nothing about the details of how the neurological substrates are involved in various moods. And although we know more than we did 50 years ago, we still don't know how mood originates and how it's translated from neural activity to conscious awareness. Nevertheless, we can expect that the parts of the brain involved in mood play some role in mood disorders.

Mood Disorders

The psychiatric classification for mood disorders encompasses a wide variety of conditions, ranging from the extreme and perhaps most socially visible form identified earlier in this chapter—bipolar disorder—to mild mood disorders whose mix of symptoms befuddle both psychiatrists and patients. Indeed, in this latter category especially, the line between "disorder" and an ordinary, if unhappy, mood state is difficult to draw, the position of the dividing line dependent on who makes the diagnosis and in which culture and what era.

The most common mood disorder in America is major depressive disorder, and we have concrete evidence relating it to the fetal environment—which is why it's discussed here.

The term "major depressive disorder" refers to a distinct change of mood in an individual, a new mind state characterized by sadness or irritability and loss of interest or pleasure in activities that one once enjoyed, and accompanied by three or four of the following symptoms: physiological changes, such as disturbances in sleep, appetite, and sexual desire; slowing of speech and action; inability to concentrate, think clearly, or make decisions; feelings of worthlessness or inappropriate guilt; and suicidal thoughts. A clinical diagnosis requires that five or more of these symptoms last more than 2 weeks and render the individual essentially dysfunctional and that they are not otherwise associated with other conditions (for example, substance abuse, a medical condition, or side effects of a medication).

The lifetime prevalence of major depressive disorder is 10–25 percent for women and 5–12 percent for men, as compared to 6 percent for minor depressive (dysthymic) disorder.

Major depression is a complex disorder with a highly variable course and an inconsistent response to treatment—and no established biogenic or psychogenic mechanism.[55] We really don't understand it, we don't know how to treat it in all cases, and we don't know what causes it. We have only some intriguing correlations.

In the United Kingdom, for example, women whose birth weight was less than 6.6 pounds have an increased risk of depression at age 26 years. In men, those born weighing less than 5.5 pounds are more likely to be psychologically distressed at age 16 years and to report a history of depression at age 26 years. It's apparent that impaired neurodevelopment during fetal life may increase susceptibility to depression.[56] This is consistent with the idea that fetal programming may affect hormones and neurotransmitter secretions that influence later mental and physical health.[57]

In Australia, among adolescents 15–17 years old, the odds for a depressive disorder are 11-fold higher for those born full-term with a low birth weight (less than 5.5 pounds) or for those born prematurely (earlier than 3 weeks before normal term). Results are the same for men and women.[58]

Clinical Disorders of Eating Behavior

The two major eating disorders are anorexia nervosa and bulimia nervosa, and there's nothing modern about them—they've been recognized for centuries, at least since 1694.[59]

The major symptom and consequence of anorexia nervosa is emaciation resulting from an obsessive pursuit of thinness. The body image, the mental idea a person has of his or her body, is distorted. There's an intense fear of gaining weight. One consequence in females is a cessation of menstruation. Weight loss occurs by extreme dieting, by excessive exercise, by self-induced vomiting, and by the use of laxatives, diuretics, or enemas.

Eating disorders are an example of how higher order dysfunction can overrule physiological homeostasis. For the anorexic individual, the intense and irrational fear of gaining weight overwhelms the need to alleviate the sensation of hunger, and anorexics literally starve themselves to the point of severe malnutrition and usually death if not treated. Anorexia nervosa has the highest mortality rate of any mental disorder. Its prevalence is about 0.3 to 1 percent. The peak age of onset is 15–19 years, and familial correlations are high.[60] Its occurrence in females is about 10 times that in males.[61]

Bulemics also have a strong concern about their weight, but during irrational binge eating that concern is muted. During an eating binge, a bulemic can consume as much as 11,000 calories in a few hours, then vomit after the binge or compensate by severe dieting.[62]

A statistically significant season-of-birth variation occurs in patients with eating disorders, the peak season occurring in May, with a peak for younger patients (early teens) in March.[63] As I've already noted, season of birth has also been correlated with other mental disorders, particularly schizophrenia for winter births, and the first quarter of the year for affective mental disorders, such as bipolar disorder and major depressive disorder.[64]

Pregnancy complications, such as maternal anemia, maternal diabetes, and placental dysfunction are predictors of the development of anorexia nervosa in offspring.[65] Maternal smoking during pregnancy is associated with bulimia nervosa in offspring, but not with anorexia nervosa.[66]

The risk of developing anorexia nervosa in female twins is higher than in male twins, as expected from the female/male ratio among adults in general. But an exception is the male member of an opposite-sex twin pair. Such males have a higher risk of developing anorexia nervosa when compared with other males—the risk at a level not significantly different from that of their twin sisters.[67] The most reasonable conclusion is that the male twin in an opposite-sex twin pair is exposed to a hormone or hormones dependent on the presence of a female fetus in the womb—and the exposure is a risk factor for anorexia nervosa.

Of course we need to be careful how we interpret findings that give us correlations between prenatal impact on the developing fetus and such postnatal psychological consequences as temperament and mood or disorders like depression, eating disorders, or any other mental illness. These connections remain suggestive only, markers for further research. That's the downside. The upside is that given the complexity of the human brain and behavior, there's still much value even in a suggestion about how things work and about the origins, prenatal and otherwise, of dysfunctional childhood or adult behavior. For that suggestion may ultimately lead toward the prevention of psychiatric disorders that, no matter how they're labeled, bring untold suffering to countless people.

TEN

MUCH ADO ABOUT IQ

This chapter is about the effects of the prenatal environment on childhood and adult intelligence quotient (IQ), but it needs to be more than that, since once the prenatal environment is admitted as an important determinant of postnatal behavior and cognitive performance, some old hereditarian ideas must be abandoned. Abandonment of these old ideas will be a critical step forward in the twenty-first century, and I want to outline here the reasons for the shift among researchers. The problem for the general reader is that although the new view of IQ involves concepts in statistics and psychometrics, I have committed myself to tell the story without graphs or equations. I do hope this works for the general reader, but if in a few places the discussion seems too technical, I advise the reader to just focus on the plain English. As with other chapters, the endnotes give many sources for further reading and explications of technical points.

MELTING POTS AND IQ TESTS

One of my earliest memories of elementary school is of being instructed that America is a "melting pot"—a place where people of different cultural and ethnic heritages live together in harmony. The image I always had in my mind was of a large metal pot with a fire under it and people dumped into it to melt into a human soup.

Americans like to think of themselves as members of a noble motley tribe that accepts the idea that human differences are unimportant. In fact, Americans invented most of the techniques used to measure and describe differences between people. The French did invent the first IQ test (the Binet–Simon Scale, 1905) to determine which children needed special attention in the classroom. But Americans quickly grabbed at the French IQ test (ultimately revising it as the Stanford–Binet Intelligence Scale), devised a few more such tests, and used them to determine which people were smarter than other people, and especially which groups of Americans were smarter than other groups of Americans. Before long the federal government was using IQ tests to "measure" which foreign groups could be allowed in as immigrants without contaminating the American gene pool, and state governments were using IQ tests to determine which children should be castrated because they were too dumb or too unruly or masturbated too much to have their genes passed to another generation. Some have since dubbed this policy "negative eugenics."

We don't use IQ tests as a reason to castrate people anymore, but the history of IQ tests in America remains as ugly as you might imagine. It's a business, as much of a business as selling cigarettes, and almost as pernicious. IQ testing is now so ingrained in American education and business and in media myths about differences in intelligence between ethnic groups, that like tobacco there's no evident method for getting it out of our culture.

Meanwhile, the current pervasive myth about IQ is that it's an innate and inherited trait, determined by genes and with no or only minor influences by the individual's environment. The myth is substantially based on a set of mistaken ideas,[1] one of which imagines that babies pop out as immaculate gestations and that individual IQ is unaffected by the prenatal environment.

The origin of most myths about IQ is the field of American behavior genetics. Psychometrics refers to the construction and application of psychological tests (for example, the IQ test), while behavior genetics is a field of study concerned with the genetic basis of animal and human behavior. Not all behavioral geneticists use psychometric tools, and not all psychometricians (people who construct and apply psychological tests) are behavioral geneticists. Controversies are sharp: there are controversies about the construction and use of IQ tests by psychometricians and controversies about theoretical interpretations of test-score distributions by behavior geneticists.

An important problem in the field of behavior genetics is that too many behavior geneticists simply assume that genetic variability underlies individual differences in cognitive ability and set themselves the task of finding correlations consistent with the assumption.[2] Some behavior geneticists are even more expansive in their claims about genetic determinism: They conclude that "there is now strong evidence that virtually all individual psychological differences, when reliably measured, are moderately to substantially heritable."[3]

And what are these so-called reliable measures? One is based on a variable called "general intelligence," also known as the "g-factor" or the "Spearman g-factor," after Charles Spearman, the British psychologist and psychometrician who proposed it in 1904 as part of a larger theory of intelligence. He defined it as the basic ability that underlies the performance of different intellectual tasks. Genetic-determinist psychometricians have promoted it in their various pronouncements about intelligence and its place in IQ tests ever since.[4] But the g-factor is not a tangible biological variable. Instead, it's a statistical construct with no clear biological referent. After more than 100 years since its introduction, this construct has contributed nothing of importance to our understanding of human behavior and human differences—and certainly nothing of importance to neuroscience.

The father of psychometrics was the Englishman Francis Galton, who contributed an enormously silly idea that in time became transmogrified into the justifications of the ugly eugenics movement that spawned public policies of the sort just mentioned. Galton's idea fed into the myths of inheritance that numerous psychometricians later embraced with a vengeance. One of them, arriving a generation after Galton, was the British psychometrician Cyril Burt, who between 1943 and 1966 was hailed for his studies of twins in England, studies that supposedly proved intelligence is an inherited trait. By the 1970s the game was up when Burt's data were demonstrated to be faked.[5] As we will see later, even more recent twin studies have been problematic in design and their interpretation controversial.

Psychometrics, the techniques of mental measurement, has indeed had a sorry history. The result of it all is has been a media carnival that has fomented a great deal of public confusion about the origins of human behavior. A perfidious tent in the carnival is the use of psychometrics to promote wrong ideas about group differences in intelligence and behavior and racist ideas of white supremacy. The ideas of behavior genetics applied to cognitive performance keep getting reburied only to rise again.[6] Complaints from the physical sciences about the statistical methods used in behavior genetics have been continually ignored.

THE "HERITABILITY" CALCULATION

We need some terminology here. "Variance" is a term in descriptive statistics, a measure of the degree of variability of a distribution of data. In the present context, it's the variance of IQ test scores that's at issue. Behavior genetics uses psychometrics (for example, IQ tests) to measure a cognitive performance (for example, IQ) under various conditions, and then makes an analysis of results to derive the "heritability" (genetic contribution to variance) of that cognitive performance (for example, the heritability of IQ). In plain language, what the behavior geneticist wants is an estimate of how much of IQ variance is due to differences in inherited genes. The scientific controversy is whether the behavior geneticist ever gets what he wants.

Heritability refers not to a trait in isolation but to a trait in a specific population exposed to a specific set of environments—which means the trait may have different heritabilities in different populations in different environments.[7] This is axiomatic in the biological sciences, but often ignored in the behavioral sciences.

There's another important difference in the approach to heritability between biology and behavioral genetics. Biologists are usually interested in how well a trait will respond to selective breeding, while behavioral geneticists have been focused on how much of a trait is determined by genes.[8]

The major problem in behavior genetics is that a crude approximation must be used to get an estimate of how much genes contribute to the variance of any given human trait.[9] The calculated quantities that are derived from such approximations are unfortunately useless to understand mechanisms or as guides for public policy. As far as I know, there has never been a refutation of that idea, and it constitutes the central complaint of biologists against the application of behavior genetics in cognitive science.

So what is the "crude approximation" here? In general, the "heritability" calculation of behavioral geneticists assumes a linearization of an ill-defined nonlinear function relating phenotype variance to gene-derived variance, environment-derived variance, and an assortment of partial interactive variances from various sources. The linear approximation considers the total phenotypic variance to be simply the sum of variance due to heredity plus the variance due to environment. Heritability is the proportion of phenotypic variance due to genes, so according to the linear model heritability plus the proportion of variance due to environment must always be equal to 1.0.

In science, linearizations are notorious for distorting reality. The fact that we don't know the nature or magnitude of the partial variances in IQ test-score analysis is not an excuse for using an amputated linear equation and then drawing momentous conclusions (for example, about the heritability of intelligence and behavior) from the results. The rule in science is that extreme care must be taken in interpreting first-order approximations. The scheme used in behavior genetics is a first-order linear approximation that discards all nonlinear terms, including terms involving interactions between genes and environment. The scheme offers no evidentiary estimate of the magnitudes of the terms discarded. If any discarded term is of the magnitude of any retained term, the entire vaudeville act falls through the floor. The evidence in this book is in fact a demonstration that many of these interaction terms may be quite large.[10] I'll return to some of these points later with a thought-experiment illustration of how such psychometric calculations for IQ can be misleading.

A DIGRESSION ABOUT BRAIN SIZE AND IQ

Since we already know that changes in the fetal environment can cause severe changes in the development of the fetal brain and nervous system, one question that pops up is whether mere brain size can explain IQ differences—brain size either determined by genes alone or brain size determined by gene–environment interactions during gestation.

People who promote the idea that differences in IQ reflect differences in inherited genes usually also promote the idea that IQ differences are based on differences in the size of the brain. But there is absolutely no evidence that the total size of the adult brain is of primary importance in intelligence.

Total adult brain volume accounts for only 16 percent of the variance in IQ scores.[11] Brain size does not predict general cognitive ability within families.[12] In addition, mentally retarded children often have larger brains than ordinary children. Women tend to have smaller brains than men. Albert Einstein had a substantially smaller brain than most men his size.[13] It's the connections of the brain, particularly the connections in the cerebral cortex, that determine cognitive performance.[14] The total volume of the adult brain tells us nothing about how nerve cells are connected to each other. Touting brain volume as an indicator of intelligence is like proposing the size of a computer as a predictor of its computing power. Years ago, many researchers had in their labs expensive massive boxes with hardly any memory and a capacity to do next to nothing important—mentally challenged "personal" computers. Computing power and performance depend on components, software, and internal connections, and not on volume of the apparatus.

So differences between people in the sizes of their brains can tell us nothing about the origins of their differences in IQ—and have little relevance for any questions we ask about the fetal environment and intelligence.

PRENATAL ENVIRONMENT AND IQ

Cognitive performance is brain performance. In general, any prenatal impact that affects the fetal developing brain has the potential to affect childhood and adult cognitive performance. There is no biological reason why this should not be so.

The clearest refutation of the idea advanced by hereditarian behavior geneticists that the prenatal environment is of only small consequence for childhood and adult IQ is the research on known IQ effects of prenatal exposure to certain neurotoxins. Any argument that this neurotoxic impact is extreme, rare, and therefore irrelevant is unsound. In addition, every known neurotoxic effect confirms the possible prenatal impact of other environmental agents not yet studied, and we do know there are literally hundreds of such neurotoxins already dispersed in the environment.[15]

I've reviewed in previous chapters a number of neurotoxins that we know affect IQ. For example:

- Lead can apparently affect IQ without any threshold. The current CDC "level of concern" for lead is 10 units, but at 10 units the IQ deficit is already 7.4 points compared with a blood lead level of 1 unit.

- In the United States, approximately half a million children are born each year with enough mercury in their bodies to cause IQ deficits ranging from 1 to 24 points. The downward shift in IQ resulting from industrial emission of mercury is more than 1500 excess cases of mental retardation each year.

- Common high levels of polychlorinated biphenyls (PCBs) in fetuses and children are associated with IQ deficits in many places in the industrialized world.

- Consumption of an average of two or more alcoholic drinks a day by a pregnant woman will produce an IQ deficit in her child of about 7 points. Even less than one drink a day during pregnancy causes a deficit in offspring cognitive performance. The current estimate is that 15–25 percent of pregnant women drink alcohol during pregnancy.

The major problem concerning the impact of environmental neurotoxins on fetal brain development is not the absence of evidence but the absence of research on the many new chemicals in our midst. Regardless, the evidence that we do have tells us that simplistic genetic-determinist arguments about IQ are dubious.

TWIN STUDIES AND HERITABILITY OF IQ

The major body of evidence used by behavior geneticists who propose inherited determinants of intelligence consists of studies of the IQ of identical and fraternal twins either reared apart or reared together, sometimes with corollary studies of the IQs of their parents and siblings. Many psychologists outside this group complain about the verifiability of the data, including complaints that the raw data for some major studies have never been published and are unavailable on request with an excuse of privacy concerns. I don't want to get involved here in this tangle, since my view is that there are more fundamental problems with twin studies. I think this chapter makes it clear that twin studies provide no refutation of the idea that the fetal environment may be of tremendous importance in shaping IQ.

The most important fundamental problem is that human twin studies of IQ heritability are too crude to generate meaningful quantitative results.[16] And the qualitative results that are generated do not unequivocally demonstrate that IQ is determined primarily by inherited genes, although many newspaper and magazine articles declare just the opposite, with studies of twins often waved as proof of genetic determinism.

A second fundamental problem is that twin research involving monozygotic (MZ) (identical) twins in the study of ordinary and pathological behavior can confirm environmental involvement, since if the inherited genomes are identical, only environment remains as a possible contribution to variance, but such research can never confirm genetic involvement, since similarities or differences in human environments cannot be established with certainty and quantified. The reason for this is simple: humans are a complex interacting species, and no two human environments are ever identical. Without identical environments, there is really no way to eliminate or quantify environment as a contribution to variance in the simple designs used in twin studies. On the other hand, MZ twin studies can be extremely useful to provide insights of no interest to hereditarians—such studies can be used to establish discordance in identical twins.

The existence of discordance in behavior or IQ in an identical twin pair suggests either prenatal or postnatal environmental influences are involved in the differences—and if we're smart enough we may be able to pin down the source of influence.

Like any other fetus, fetal twins have a biology. Attention to the variations in fetal brain development of twins is essential to any understanding of the implications of twin research for analysis of behavior. In the United States, the mean gestational age of twins is 3.5 weeks less than that of singletons, a substantial difference.[17] Short gestation is associated with poor birth outcomes, long-term morbidity, and higher infant mortality rates. Moreover, twins have substantially lower IQs than singletons in the same family, an effect attributed to prenatal developmental differences.[18] It's thus a mistake to consider twins as a simple and important "natural" experiment in genome identity, especially since it's gene expression that rules and not genes.

Another problem is whether MZ twins are actually "identical." In fact, by the time gestation is finished, most MZ twins are not identical. There can be major discordances for birth weight, genetic disease, and congenital anomalies.[19] The twinning process may produce an unequal allocation of cells with important consequences for the brain development of each twin. It's unfortunate that the authors of some major psychometric twin studies have refused to publish raw data that might reveal important physiological and psychological discordances. Their arguments for not doing so—because it would compromise the privacy of the participants in these studies—seem inadequate, since there are many ways to guard privacy of subjects in research. If raw data must be hidden from other researchers, maybe it would be better not to publish the results of such studies.

The known differences between MZ twins can be substantial. Let's consider the three basic uterine "housing" arrangements for MZ twins.[20] Apart from the placenta, there are two important structures, both membranes, associated with the developing fetus. The amnion is a thin membrane that forms a closed sac around the fetus and contains the amniotic fluid. The chorion is a thick outer membrane involved in the formation of the placenta.

In one housing arrangement, the MZ twins have separate amnions, separate chorions, and separate placentas. This occurs when the twins derive from a splitting at the two-cell stage.

In another arrangement, the MZ twins have separate amnions, a common chorion, and a common placenta. This monochorionic arrangement occurs when the twins derive from a splitting of the inner cell mass of an early blastocyst.

In a third arrangement, the MZ twins have a common amnion, a common chorion, and a common placenta. This occurs when the twins derive from a later splitting that yields two embryos from one blastocyst inner cell mass.

Lower birth weight of an MZ twin compared to the cotwin is associated with lower adult intelligence compared to the cotwin.[21] This is a fetal-environment effect that can result from an MZ transfusion syndrome, which occurs only with monochorionic MZ twins. The syndrome is a consequence of vascular fusions that cause an unbalanced blood supply between the twins. It is usually fatal for both, but when the twins survive, one twin usually has a lower birth weight. The transfusion syndrome occurs in 5–15 percent of monochorionic MZ pregnancies.[22]

Indeed, as research in human genetics continues, it has become clear to the medical community that the identity of "identical" twins is not as useful as previously thought, and that it may be more important to understand discordances between MZ twins than to understand their similarities.[23]

Despite all of this, identical twins are usually more similar than different. But the most important biological characteristic of MZ twins, in the context of the questions asked in this book, is that for the most part they share a single maternal nutrient environment, and the mother bearing the twins is the single filter of the external environment for both twins. Hereditarians base their arguments for the idea that genes rule IQ primarily on studies that show high pair correlations in IQ scores for identical twins reared apart.[24] The argument is that if environment rules, why would the correlations be so high? The "environment" they look at is the environment after birth—supposedly different environments in identical twins reared apart. Acceptance of the importance of the fetal environment completely undermines the hereditarian argument, since with identical twins the fetal environments are usually more alike than different, and much more alike than the fetal environments of fraternal twins. So the similarity of fetal environments in identical twins can explain the high IQ pair correlations without recourse to the idea of inherited genes determining cognitive performance.

Some have suggested that studies of MZ twins with shared or nonshared chorions are evidence that the fetal environment is not important.[25] Since the idea was recently presented in popular media, it needs to be discussed and refuted. The basis for the idea is that whether MZ twins share or do not share a chorion seems to have little effect on IQ correlations of twins in a pair.[26] The fallacy here is that the chorion does not fully define the fetal environment. Do dichorionic MZ twins have different fetal environments? It's unlikely. An analogy would be a claim that two children who live in adjacent buildings in an urban ghetto have different environments. They do if you define "environment" in terms of the building one lives in. Both monochorionic and dichorionic MZ twins essentially share the same environment—the maternal uterus. The chorion argument for the unimportance of the fetal environment is not sensible.

One key assumption of many twin studies is that both MZ and dizygotic (DZ, or fraternal) twins reared in the same home have been exposed to the same environments, so that MZ twins and DZ twins are equally correlated for their exposure to relevant postnatal environmental influences.[27] According to genetic determinists, if the postnatal environment were the most important differentiating factor in IQ correlations, MZ twins and DZ twins should both have about the same correlations for IQ: if the environment rules, and the environments are the same, the correlations should be the same. But MZ twins, who share the same genome, always have much higher IQ correlations than DZ twins, who inherit a random half of each parent's genes. If environment rules, why should MZ twin correlations always be much higher than DZ twin correlations? Don't the results mean that genes rule?

Definitely not. The "similar" environments in these studies is the postnatal environment. The prenatal environment is never considered in the analysis. For twins reared in similar postnatal environments, we don't know if the higher twin-pair IQ correlations for MZ twins are due to identity of genes or similarity of fetal environments. It is certainly possible that the fetal environments (including reactions to impacts) of identical twins are always more similar than the fetal environments of fraternal twins. We don't know enough yet about human twin biology to discount that possibility.

In contrast to the older arguments for IQ as a heritable trait is the newer evidence that fetal environment effects account for 20 percent of the covariance between twins and 5 percent between siblings.[28] The consequence is that, according to the linear variance approximation made by behavior geneticists, heritability of IQ must be less than 50 percent. Even assuming the validity of the linear heritability calculation, a heritability of less than 50 percent does not allow any suggestion that IQ is determined primarily by inherited genes or that IQ variance is determined primarily by differences in inherited genes. In plain English, during the past decade it has become evident that even with the crude approximate methods of behavior genetics, the old idea that most of the differences in IQ are explained by heredity is incorrect: the percentage is much less than 50 percent. So according to the model used by behavior genetics, it's the environment that rules, not genes.

Too many problematic explanations, including some based on fallacy, have been promoted by hereditarians. Consider the apparent increase in the so-called heritability of IQ as a twin cohort ages. Genetic determinists often say this means the role of genes in determining IQ becomes more important with age.[29] Not so. As people grow older in any society, the cultural differences between them tend to diminish as a result of common experience. MZ twins 40 years of age in the same socioeconomic class have enough shared environment and shared culture to substantially increase any calculated "heritability" fraction of variance in IQ over what it was when the cohort was 10 years old. This results from the simple mathematical fact that as differences in individual environments are reduced, the contributions of environment to variance are reduced. When individual environments are identical, "heritability" of IQ would be 1.0—all the variance would be due to heredity because according to the analytic paradigm there would be no other source of variance. Fallacies result when analysis of variance is mistakenly segued into analysis of causality.[30]

The general constraint on the interpretation of all twin studies is that differential effects of culture, family, and prenatal environments, and even the effects of their physical appearance on the correlated experience of twins, all affect twin development.[31]

Let's consider a related question: What does a twin pair-correlation coefficient tell us anyway? If the correlation is high, does that mean the cognitive performance of the twins is very similar? Well, sometimes yes and sometimes no. I want to demonstrate here how it's possible for the answer to be no. We will do a thought experiment, plug in some numbers, do the statistics and see what we get.[32] My objective is to demonstrate that a high pair correlation can mask important differences, so that without raw data the correlation numbers may produce deception.

In our thought experiment, we start out with 10 pairs of MZ twins. We might consider 1000 pairs—the result will not change substantially. These twins (all males) are born into ordinary middle class families in a Midwestern suburb. Let's assume that the cotwins of each pair have identical IQs at the age of 2. The twins are then separated at that age, one twin remaining at home and the other twin moved into a lower-class family in an urban ghetto. We allow 30 years to go by, during which time the twins are totally out of contact with each other. Then we give standard IQ tests (for example, the Stanford–Binet) to all the twins.

Such is our thought experiment. We will imagine a set of results for the twins who remained home and for the twins who moved into ghetto foster homes. Here are the twin pairs at age 32, rank-ordered in IQ according to the original-home twin:

Original-Home Twin IQ/Foster-Home Twin IQ

90/72
94/80
100/85
106/93
108/94
118/103

120/100

122/110

130/118

137/117

What has evidently happened in this thought experiment is that if the twin pairs started out at the age of 2 years with the same IQ, moving one twin of each pair into a ghetto foster home has resulted in an IQ drop of 12–20 points—maybe enough points to turn a stock exchange floor trader into a stock exchange floor sweeper.

At 2 years of age the twins may have had the same IQ due to genes or to the fact that they came out of the same uterus or due to some combination of variables—we don't know. What we do know is that every foster-home twin in this thought experiment has evidently experienced a large IQ deficit produced by the change in postnatal environment, the one that is probably the most important developmentally.

But now let's take the IQ data from our thought experiment and calculate the standard Pearson correlation coefficient (named after the classic statistician Karl Pearson). You may be surprised at the result. What we find is a pair correlation of 0.98, just a shade under what would be called a perfect correlation.

We thus have a high correlation between the separated twins in spite of substantial differences between cotwins in IQ scores. If we're not told the raw scores, this extremely high correlation would seem to suggest that MZ twins inherit their IQ. But the circumstances of our thought experiment and the actual raw scores suggest nothing of the sort. They tell us something quite different instead: that it's possible to have a high correlation between two arrays even if the absolute values of the elements in the arrays differ substantially. The reason for this is that a correlation coefficient is not a measure of similarity of magnitudes but merely a measure of similarity of variation. For example, two straight lines with the same slope have the highest possible correlation coefficient, 1.0, even if their corresponding points differ by orders of magnitude.

The correlation coefficient by itself not only tells us nothing about cause and effect, about what "determines" what, but in the case of our thought experiment the correlation hides the most important part of the experiment: the fact that moving one twin from each pair from a middle-class home into a lower-class home can cause an IQ deficit of anywhere from 12 to 20 points.

Of course no one has done such an experiment. The calculated correlation is correct, but I cooked the raw data. It's a "thought" experiment to make a point about twin studies and IQ: we need to be careful about how we interpret IQ twin correlations—especially when raw scores are not published or even available on request.

Related to twin studies are so-called adoption studies, which examine the IQ scores of children who were adopted early in life and compare their scores with the IQ scores of both their natural and their adoptive mothers. The adopted child and its adoptive mother supposedly share environment but not genes, while the adopted child and its natural mother share genes but not environment. The idea is to examine IQ correlations to see what apparently contributes to variance. It is assumed that whatever variance is not "environmental" must be "genetic" in origin. One problem is that there are too many ways to interpret the data.[33] Different studies produce conflicting results, depending on design, subjects, and the biases of researchers. For example, it's difficult to demonstrate that an adopted child and its natural mother do indeed have different environments, or that an adopted child and its adopted mother indeed have the same environment. Like twin studies, adoption studies do not provide sufficient evidence for genetic determinism of IQ.

NEW QUESTIONS ABOUT FETAL ENVIRONMENT, HEREDITY, AND IQ

The science of human behavior and its origins is of such enormous importance that as long as unfettered science exists new facts and ideas ultimately come along to move us forward in our understanding. During the past 30 years, American hereditarians have sold the public the idea that inherited genes contribute 40–70 percent of the variance in IQ and that IQ is essentially an unchangeable behavioral trait. This view has been shattered, first by the growing recognition of the importance of prenatal environment, second by new evidence that socioeconomic class

determines the contribution of genetic heredity and environment to total IQ variance, and third by a phenomenon called
the "Flynn effect" that involves a substantial increase in IQ
scores with each generation, an increase apparently in all groups
and nationalities.

It's characteristic of American media that none of this new evidence has received as much attention as the myth that genes rule.

In this subsection I will deal with the effects on IQ of socioeconomic class (the Turkheimer study), and in the next subsection with the Flynn Effect. These are two important new
developments that concern individual IQ and need to be discussed in detail. This whole section is the first part of myth-
debunking. The second part appears in the final chapter and
deals with an array of other myths that are more general.

The Turkheimer Study

The new socioeconomic evidence first appeared in 2003,[34] in
what has been called the Turkheimer study, named after the lead
author of the article that reported the study's findings. What the
study shows is that in poor families nearly all the variance in IQ
is accounted for by a combination of fetal and other environments, and the contribution of genes to variance of IQ is close to
zero. The result is almost the reverse in affluent families—nearly
all the variance is accounted for by apparent heredity.

At first glance it's a puzzle. But what is involved here is variance, not causality, so maybe the puzzle is not that formidable.

The most reasonable explanation of the results is that among poor children genetic differences contribute almost nothing to the measured IQ variance because environmental damage, both fetal and postnatal, overwhelms all other variables in accounting for IQ variation. In contrast, in the middle and upper classes, in which fetal and postnatal damage to the nervous system is much reduced and hardly variable from one family to the next, genetic differences account for most of the variation in IQ.

In the next chapter I will discuss the impact of socioeconomic factors on the fetal environment. Such factors are not minor, they're numerous, and they vary widely in character and intensity. It's not a surprise that these factors, most prevalent in lower socioeconomic classes, will overshadow any genotype contribution to IQ variance.

The Flynn Effect

If we're to evaluate the implications of the effects of fetal environment on IQ, we need to understand the limitations of the IQ test. The construction of an IQ test involves a normalization process. Items in the test are selected and included so that a sample population will produce a symmetrical bell-shaped distribution of test scores with a mean and median of 100 points and a standard deviation of 15 points. The test is renormalized at intervals of 20 or 25 years to keep the parameters of the distribution curve the same: mean and median of 100 points and standard deviation of 15 points.

In the 1980s it was discovered that the renormalizations of the Stanford–Binet IQ test, ongoing since its introduction in America in 1932, and the renormalizations of the other major IQ test, the Wechsler Adult Intelligence Scale (WAIS), ongoing since 1953, had been hiding the fact that the performance of the American population on IQ tests had been increasing about 3 points per decade or approximately 10 points per generation.

This is the Flynn effect, named after James R. Flynn, the New Zealand psychologist who first brought the phenomenon to everyone's attention.[35] Within a few years it was further discovered that the effect is also occurring in 14 nations around the world.[36] The number of nations has now grown to 23.[37]

The 23 nations that have shown massive gains in IQ are as follows: Argentina, Australia, Austria, Belgium, Brazil, Canada, Denmark, Dominica, Estonia, France, Germany, Great Britain, India, Israel, Japan, Kenya, Netherlands, New Zealand, Norway, Spain, Sweden, Switzerland, and United States.

The first study to document the Flynn effect in a rural area of a developing country (Kenya) was reported in 2003.[38] From 1984 to 1998, the Raven's IQ score in that rural district increased 4.5 points. The Raven's (known as Raven's Standard Progressive Matrices) is a widely-used, mainly culture-free, IQ test involving analogies of geometric shapes. Results from a study on the Caribbean island of Dominica were reported in 2005.[39] In this study a Flynn-effect increase in Raven's IQ score of 17 to 19 points apparently occurred over a period of 37 years.

It's probable that sooner or later every nation in the world will be included in the list of countries where massive gains in IQ are apparent from one generation to the next. No one knows why this is happening. There are speculations, including some by Flynn himself, but the Flynn effect remains unexplained.[40] What seems clear, however, is that the effect must involve one or more strong environmental components. There is no known way for human genetic components to undergo Darwinian natural selection in only one or a few generations. In addition, the global scale of the effect in fact precludes any explanation restricted to genetic changes alone. Environment may be acting alone or affecting gene expression—but environmental changes must play a role in the increases in IQ that have occurred.

What environmental changes? One of the most reasonable guesses about the origin of the Flynn effect is that it may be due to global improvements in nutrition, prenatal care, and maternal education. Another contributing factor may be the detoxification of the maternal environment, and with it the detoxification of the fetal environment, that has occurred in many places. In short, perhaps IQ is increasing because damage to the developing fetal brain is being reduced by education and public policy in those places. Or perhaps the gains might have multiple causes and more than one cause operating in a single nation.

Whatever the reasons for it, everyone agrees the Flynn effect is a real phenomenon. And some of its apparent implications are startling. If IQs have been rising with each generation, then earlier generations had lower IQs. For example, it's estimated that 70 percent of late nineteenth century Britons had an IQ of 75 when scored in current terms.[41] The Flynn-effect consequence after analysis of IQ tests given by the Dutch military is that Dutch men in 1952 had a mean IQ of 79 when scored against 1982 norms—which implies that the people of the Netherlands in the 1950s had an average intelligence close to mental retardation by current standards. If the comparisons are done in reverse, the implication is that substantial fractions of whole populations have been moving into the high-intelligence range during the past 50 years.

Truly, it seems that something is rotten in IQ-land.

In short, the existence of the Flynn effect seriously weakens the usefulness of IQ tests. For example, IQ tests are used in public policy directives to identify children with mental retardation. But many children who are labeled mentally retarded by IQ test performance today would not have been so labeled 50 years ago. A child with an IQ tag of 70 today would have an IQ score 50 years ago of 86.5 in its own cohort and be out of the mentally retarded group. In parallel, the labeling of some criminals as mentally retarded, crucial for sentencing purposes, can depend on whether an older or newer version of an IQ test is used to determine the label.[42]

What is clear is that we know too little about what IQ tests measure. Whatever it is that they measure, the effects of environment are evidently more powerful than many people suspected. For a hundred years now the American public has been bombarded with theoretically unsound statistical demonstrations that individual differences in IQ are determined primarily by heredity. It's a canard that deserves to be drummed out of discourse.

PART III

MISERY FOR ALL SEASONS

ELEVEN

Culture, Poverty, and Fetal Destruction

I like wine. I like Italian wine better than any other wine, maybe because I once lived in Italy. In those days I was certain Italian wine was nature's way of softening life's brutality. Now I'm older and less gullible. Things that soften nature's brutality are sometimes brutal themselves.

To many people, Italy means wine. To Italians, wine is paramount, the wines and the vineyards that produce the wines are part of the Italian national soul. The cultural traditions of Italy have strong influences on its people. Drinking habits are distinctive. Binge drinking is rare except among the young and in large cities. In rural Italy, the custom is to drink moderate amounts of wine with meals every day. Italians in Italy who don't drink wine are rare. You don't drink wine? they say. They shake their heads in wonder.

One of the oldest parts of Italy is Rome and the area around Rome, the region called Lazio. In the region of Lazio, 62 percent of women drink alcohol before pregnancy and 53 percent during pregnancy; 12 percent of pregnant women drink seven or more drinks a week.[1] Some Italian women who ordinarily don't drink regularly start drinking during pregnancy because of a popular belief in Italy that moderate alcohol has benevolent effects during pregnancy.

The result? Currently, 3.5 percent of the children in Lazio have the markers of fetal alcohol spectrum disorder (FASD).[2] That's the highest prevalence of FASD in the Western world and more than three times the prevalence in the United States.

It's true that nearly everyone in the world likes wine of one kind or another. It's also true that wine mangles the bodies and brains of unborn children.

In this chapter, I want to give examples of how various cultural factors, in general, and poverty, in particular, impact the fetal environment and through that impact shape later behavior and IQ.

But what is "culture"? The term appears everywhere, in newspapers, magazines, broadcast media, and scholarly articles. For professionals in the social sciences, culture is defined as "the sum total of the ideas, beliefs, customs, values, knowledge, and material artefacts that are handed down from one generation to the next in a society."[3] So culture is essentially the human world that the individual finds himself in. Everyone of us is at the same time in a local culture, a national culture, and a global culture—and these cultures play a role in how we think, how we behave, how we respond to whatever is happening around us. And the different cultures (local, national, and global) affect different people in different ways. The pregnant women of rural villages around Rome are more influenced by local culture than by national or global culture. The amount of lead in the soil around houses in Chicago is mostly influenced by national culture—by attitudes in the mid-twentieth century about possible dangers of leaded gasoline and current national attitudes about the task of cleaning up the soil. The purity of drinking water in Topeka, Kansas is mostly influenced by local culture—by the attitudes of local industry and local politicians and of everyone else in the community. In general, the development of every individual is influenced by factors in the surrounding culture—factors that operate from conception forward, and not just after birth. No fetus develops independent of the culture around it.

The factors are diverse, ranging from caffeine consumption during pregnancy to maternal age at conception. The environmental agents with potential adverse impacts on fetal development for which there is the strongest evidence are all culture-dependent: tobacco smoke, alcohol (ethanol), cocaine, and combustion-engine carbon monoxide.[4] In this book I focus only on factors for which there is evidence known to me. What the factors all have in common is that all have heavy cultural origins (for example, the consumption of wine in rural Italy) and they all have known impacts on fetal development. Some of these factors have been discussed in other places in this book, but they need to be rounded up here in the context of cultural influences on fetal development and the consequent shaping of behavior and intelligence in children and adults.

It's important to remember that there is no single culture in America, no single way of living. What is obvious from all the numbers is that different cultures produce different risks for fetal development.

But no matter the culture, a critical phenomenon needs to be emphasized: the impact of culture on the fetal environment, one environment influencing another environment, the fetal environment translating cultural forces into biological effects that shape the development of the brain and nervous system. This connection between culture and brain development is as much anthropology as neuroscience—which makes the connection intriguing and even startling. But more important is that the connection is crucial for public policy.

CULTURE, LOW BIRTH WEIGHT, AND INFANT MORTALITY

Estimating various consequences of impacts on the fetus can be difficult. An example of uncertainties in research is the fact that Hispanic children in America have an apparent lower prevalence of autism, mental retardation, and cerebral palsy than children in other groups, but we don't know if this is an actuality or merely due to an absence of diagnosis, a lower likelihood to report such problems in surveys, or some other cause not yet recognized.[5]

In contrast, low birth weight and infant mortality (death before 1 year of age after a live birth) can be recorded and measured exactly.

At the present time, according to the World Health Organization, 35 countries have a lower infant mortality rate than the United States, including Cuba and Poland.[6]

The infant mortality rate in the United States decreased by 45.2 percent from 1980 to 2000, but the percentage of low-birth-weight births increased by 11.8 percent during that period, including a 2.4 percent increase for blacks and a 14 percent increase for whites. Infant mortality has been reduced, but the percentage of low-birth-weight children entering the population has increased. The overall black–white gap for infant mortality has widened and blacks continue to have a two- to three-fold higher risk than whites for low-birth-weight births.[7]

The increasing U.S. black–white racial disparity in infant mortality is apparently a result of improvements in the survival of white preterm and low-birth-weight infants, a consequence of improved access by whites to advances in medical technology.[8]

The American black–white difference in small-for-gestational-age (SGA) births is a pathological difference rather than a physiological difference. Here "physiological" means normal or expected, and "pathological" means a birth weight that will later be correlated with disease.

Foreign-born blacks in the United States have live birth rates, mean birth weights, SGA rates, and neonatal mortality rates intermediate between those of American-born whites and American-born blacks.[9]

But the issue of how to categorize birth weights in various populations is hotly debated, and some researchers have provided evidence and made a case for the use of ethnic-specific standards of fetal growth.[10]

In 2004 in the United States, the fetal mortality rate at 20 weeks or more of gestation for non-Hispanic black women was 2.3 times the rate for non-Hispanic white women, while the rate for Hispanic women was only slightly more (10 percent more) than the rate for non-Hispanic white women.[11]

How little we understand about cultural factors in fetal development is illustrated by the paradox of foreign-born Asian-Indian-American women of high socioeconomic status who begin prenatal care early in pregnancy and who yet have a higher incidence of low-birth-weight infants and fetal deaths compared to comparable American women of European ancestry.

In contrast, foreign-born Mexican-American women of low socioeconomic status have a lower probability (6.3 percent) of low-birth-weight births than non-Hispanic whites (7 percent) and much lower than non-Hispanic blacks (13.6 percent).[12]

Low birth weight is a measure of the health of the fetus, and infant mortality is a measure of both the health of the fetus and the impacts of the early postnatal environment. We can expect that whatever the health of the fetus might be it will affect the development of the brain and nervous system, and this development in turn will affect childhood and adult behavior. To deny these links is to deny a biological substrate for behavior, which makes no sense at all. And if we admit these links affect childhood and adult behavior, we must also admit possible affects on IQ, mental health, and criminality. Most social scientists are fully aware of these links and the complexities they involve, but unfortunately too many people everywhere still have archaic and even medieval ideas about the origins of intelligence, mental illness, and crime—ideas that usually involve separate sets of supposedly unrelated causes ranging from the toss of nature's dice to genealogy. But what the data in this book tells us is that it may be culture—local, national, and global—that's making the bed most of us lie in.

Cultural effects are complex, diverse, and not always explicit. In the following sections we consider how geography, maternal behavior, and poverty act as transforming mediators between culture and the fetal environment. What I present here is no more than a meager summary of the available data in the scientific literature; the endnotes provide sources to amplify the text. My objective is only to establish a central idea: the link between culture and fetal development.

CULTURE AND THE HAZARDS OF PLACE AND COMMUNITY

Everywhere on the planet, for people of similar socioeconomic class, urban life is more stressful than rural life. A rich family in Tokyo is certainly less stressed than a poor family in a Japanese mountain village, but a poor family in Tokyo is usually more stressed than a comparable rural family. Urban life stress is correlated with an increased prevalence of psychiatric disorders and also correlated with maternal stress during pregnancy. Women do not experience pregnancy disconnected from their surroundings, and as we'll see later, the impacts of maternal stress on the developing fetus can be dramatic.

Air pollution in cities is related to lowered birth weight and intrauterine growth restriction (IUGR). Blood levels of DNA adducts (chemicals that bind to DNA) are positively related to a risk of IUGR. The fetus apparently sequesters DNA adducts—levels of DNA adducts are typically higher in infants than in their mothers.[13]

Studies of outdoor air pollution in the Czech Republic have related sulfur dioxide, nitrous oxides, and total suspended particles in air to low birth weight.[14]

The exposure of children during fetal development and early postnatal life to air pollutants is associated with low birth weight, preterm birth, intrauterine growth restriction, congenital defects, preterm and infant mortality, decreased lung growth, increased rates of respiratory tract infections, childhood asthma, childhood behavioral problems, and neurocognitive deficits. Air pollutants involved include particulates, carbon monoxide, sulfur and nitrogen oxides, ozone, and tobacco smoke.[15]

Air pollution is not just a source of material for late-night TV jokes. It's a man-made scourge that wrecks the lives of fetuses and children.

In the United States, the greater the prevalence of low-birth-weight birth the lower the average intelligence quotient (IQ) for the various states (correlation −0.43). The correlation of low birth weight with violent crime in the states is even higher (0.58).[16]

In the context of birth-weight demographics, there is no single Hispanic group: the probability of a low-birth-weight birth (less than 2500 grams or 5.5 pounds) is 9.8 percent among Puerto-Rican-Americans and 6.3 percent among Mexican-Americans; and there is also no single non-Hispanic white group: the probability of a low-birth-weight birth is 8.8 percent among non-Hispanic whites in West Virginia and 5.6 percent among non-Hispanic whites in Washington State.[17] Since it's averages for such groups that are usually reported, it's important to remember that diversity exists. Simple geography is important: the variation in the probability of a low-birth-weight birth varies between states as much as it does between developed countries.

The low-birth-weight situation in America as a whole is far from satisfactory: if we combine U.S. statistics for all ethnic groups, there are 59 countries that have a lower probability of low-birth-weight birth than the United States.

For example, the American probability of a low-birth-weight birth (8 percent) is almost three times that in Albania (3 percent), and the probability for non-Hispanic whites (7 percent) more than twice that of Albania. The probability of a low-birth-weight birth in the state of Mississippi (11.4 percent) is greater than the probability of a low-birth-weight birth in Botswana (10 percent) and Jamaica (9 percent).[18]

The geographical differences in infant mortality within single states can be striking. Examples of significant differences between cities in the same state are infant mortality rates in Ohio. Compared to Toledo, the risk of infant mortality in Cincinnati is 53 percent higher and in Columbus 28 percent higher.[19] Even after controlling for ethnicity, maternal age, maternal education, marital status, birth order, prenatal care, gestational age, and birth weight, significant city differences persist.[20] Such differences cannot possibly be genetic in origin. They must be due to environment, and the most likely environmental factor is industrial pollution of air and water supplies that impact fetal development. Cities do differ in the kinds of industrial pollutants in their local environments. It's not the total "industrial pollution" that's important, but the specific types of pollutants to which people are exposed.

Large geographic variations in local toxins, particularly in drinking water, make it difficult to compare not only countries but also states and counties within a single country. The United States does not have a single set of standards for water purification, although the Centers for Disease Control (CDC) do make recommendations. People who drink tap water in West Virginia are not drinking the same tap water as people in Minnesota.

The usual contaminants of chlorinated drinking water are trihalomethanes (chloroform, bromoform, bromodichloromethane, chlorodibromomethane).[21] In California, women who drink more than 5 glasses per day of cold tap water are almost twice as likely as other women to have a spontaneous abortion. The most significant exposure is apparently that to bromodichloromethane.[22]

It's difficult to make a case that the variety of drinking-water standards in the United States is of any service to the public. Who but government—local, state, or federal—can be responsible for the chemical contents of public drinking water? Is pure water a luxury available only to people who can pay for it?

In addition to geographical variations in environmental pollution, lifestyle differences between urban and rural areas are an important variable. For example, the use of wood-burning fireplaces for heating and wood-burning stoves for cooking (common in some rural areas), as opposed to natural-gas or electrical utilities (common in cities), can be a problem. Wood-smoke exposure during pregnancy can lead to impaired fetal tissue growth as a consequence of hypoxia or oxidative stress resulting from smoke constituents that include carbon monoxide and particulate matter. Cooking with wood fuel during pregnancy is associated with low-birth-weight infants.[23]

Another important variable correlated with geography is infectious disease. Infection is an important worldwide source of disruptive fetal impacts. *Cytomegalovirus* (CMV) is a common culprit. The overall birth prevalence of congenital CMV infection is 0.64 percent, but the prevalence varies widely from place to place. Congenital CMV infection is associated with non-white ethnic groups and low socioeconomic status.[24] As I've pointed out earlier, this infection is endemic around the world and always a likely cause of pathogenic fetal impacts.

Infections among pregnant women related to subcultures can be significant. Among pregnant black women in Baltimore, for example, approximately 25 percent have vaginal infections (bacterial vaginosis). Risk factors for infection are vaginal douching during pregnancy and frequent intercourse during pregnancy.[25] We lack an adequate epidemiology relating such infections to fetal development.

CULTURE AND THE HAZARDS OF MATERNAL BEHAVIOR

Maternal health care during pregnancy is a major factor affecting fetal development and infant mortality. The relationship between lack of prenatal care and infant mortality is due to the relationship between inadequate prenatal care and low-birth-weight births. The main component of black–white differences in infant mortality in America is the high incidence of low birth weight among black infants.[26] Low birth weight is a measure of impacted fetal development. A central question for both epidemiology and public education policy (and part of the major focus of this book) is whether low birth weight among black infants in America and the implied fetal damage, is related to later deficits in cognitive performance. (The same question must be asked about all impacts on the fetal environment that might disadvantage fetal development in blacks more than in whites.)

In the context of maternal health care, cultural factors can be paramount. The factors that determine whether a pregnant woman seeks prenatal care are complex. For example, among American black women in some localities, if a pregnancy is unwanted, prenatal care is considered undesirable. Also, lack of transportation often results in fewer prenatal-care visits.[27] Even with free access to prenatal care, cultural factors determine how long women in various ethnic groups delay in making their first prenatal-care visit to an obstetric physician or clinic.[28]

Another variable responsive to cultural pressures is maternal age. Fetal development is not independent of maternal age during pregnancy. Culture is the essential determinant of how women view the proper age range for bearing children. In the European Union between 1980 and 1993 the mean maternal age at first birth rose to 28.6 years. Between 1991 and 2001 in the United States, the percentage of women 35 to 39 years having first births increased by 36 percent, and for women 40 to 44 years by 70 percent.[29]

A cluster of social factors are no doubt involved in such changes in demographics, but no matter what the causes, the

consequences are clear to obstetricians: women who are pregnant at advanced maternal age have an increased risk of stillbirth.
A corollary is that when there's a live birth with such women,
the spectrum of defects in fetal development can range from
subtle to subclinical to severe, including unknown effects on the
cellular anatomy and function of the brain. In the United States,
the lowest percentage of low-birth-weight births occur when
the maternal age is 25 to 29 years, with substantial increases in
percentages at ages lower or higher than that bracket.[30] Given
the correlations between low birth weight and later IQ, the children of adolescent mothers can be expected to have deficits in
cognitive performance.

Another influence of culture on individual maternal behavior involves tobacco. Apart from alcohol and nutrition, the
most common fetal impact that obviously derives from cultural
influences is probably tobacco smoking. And the most common effect of tobacco smoking during pregnancy is probably
fetal growth retardation—an "SGA infant". The term "small
for gestational age" (SGA) refers to a birth weight less than the
10th percentile for gestational age. Maternal tobacco smoking
increases the risk of delivering an SGA infant independent of
any insufficient functioning of the placenta or deficiencies in
maternal nutrition.[31]

The most common consequence of maternal tobacco smoking during pregnancy is fetal hypoxia. But the metabolites of
tobacco smoke also have an impact on fetal development.[32] In
general, the consequences of maternal tobacco smoking include

a 30 percent increase in perinatal mortality and morbidity caused by IUGR in what is now called the "fetal tobacco syndrome" (FTS), a 30 percent increase in the incidence of preterm delivery, a significant incidence of preterm rupture of the amniotic membrane, and a two- to three-fold increase in the rate of placental structural and functional problems. In addition, chronic tobacco hypoxia causes fetal hypokinesis and fetal tachycardia.[33]

Added to all this is an important postnatal effect: fetuses less than 37 weeks gestational age of mothers who smoke throughout pregnancy have a delayed response to the maternal voice, and this may have implications for later language development.[34]

Heavy tobacco smoking is especially dangerous for the fetus. Maternal smoking of more than 20 cigarettes a day is associated with maternal anemia, fetal brain hypoxia, and fetal polyglobulia (an abnormal increase in circulating red blood cells).[35]

An example of cultural differences: American-Hispanic women who are more highly acculturated to non–Hispanic-American life are more likely to smoke cigarettes during pregnancy than less acculturated Hispanic women—an illustration of how culture can shape the prenatal environmen.[36]

As might be expected, prenatal tobacco exposure during a teenage pregnancy predicts behavioral problems of offspring, particularly impulsivity and aggression. Of teenage mothers in the United States, 46 percent smoke in the first trimester of pregnancy.[37] The evidence is that the offspring of these mothers are likely to also smoke tobacco, which constitutes a transgenerational nongenetic tobacco impact.

Another culture-derived maternal behavior that affects fetal development is coffee-drinking during pregnancy. Caffeine is a drug that acts on nearly all organ systems, including the brain and it's found in almost every culture on the planet. We have substantial reasons to expect high concentrations of caffeine in fetal blood to affect fetal brain development. Do low concentrations also have an effect? The quantity of caffeine consumed in coffee and tea varies from place to place, and how much is consumed by women during pregnancy also varies. The variation includes differences between individual pregnant women, since the nausea that often accompanies pregnancy sometimes results in reduced caffeine consumption.

Drinking about 350 milligrams per day of caffeine (about 1.5 cups of coffee) evidently has no effect on the probability of miscarriage during pregnancy.[38] In contrast, at least among pregnant women in Denmark, high levels of coffee consumption are associated with an increased risk of fetal death: risk factor 1.33 for four to seven cups of coffee per day, and risk factor 1.59 for more than eight cups of coffee per day. Risk of fetal death is apparently greatest after 20 weeks of gestation.[39]

Maternal diet is almost entirely culture-dependent and constitutes a vast subject beyond the scope of this book. What we call "food" consists of consumed chemicals, and there is one chemical group that deserves maybe more attention than it gets by the media. Heterocyclic amines are mutagens, chemicals capable of altering DNA to cause gene mutations. They vary in potency and vary in their presence in the human diet.

Cooking meat produces mixtures of heterocyclic amines, and exposure to heterocyclic amines may be involved in the development of tumors, including breast, colon, and stomach tumors in humans.[40]

We don't know much about the transfer through the placenta of heterocyclic amines originating in the maternal diet. If these mutagens do cross the placenta, they can be expected to have a range of effects on gene expression and development of the nervous system. Standard diets differ markedly from one culture to another. How are these differences related to mutagens passing from mother to fetus? We don't know.

Another important dietary chemical is folate—this chemical essential in the diet for a healthy pregnancy. Although folate is now considered an essential nutrient necessary for fetal development, there are still wide variations in the way folate supplements are distributed, available, and used by different ethnic groups and cultures. During the past decade, the reported rate of knowledge by women of child-bearing age of the importance of adequate folate intake was 17 percent to 77 percent worldwide and 48 percent to 77 percent in the United States.[41] So in some places in America more than half the pregnant women are unaware of the importance of folate supplements during their pregnancy. When was the last time you saw a public service billboard or TV announcement emphasizing the importance of folate supplements? Folate dietary insufficiency of mothers during pregnancy is associated with serious neurological malformations of the fetus. Are there also subtle consequences for fetal brain development of borderline dietary intake? We don't know.

Maternal Stress

We have encountered the effects of maternal stress on the fetus earlier in this book, but any discussion of how culture affects fetal development needs to include maternal stress as an important mediator between culture and fetal development. Pregnant women are never in isolation from their community, from their culture as a whole, and any stresses caused by or promoted by these social environments are translated into maternal physiology and ultimately into fetal impacts. When the stock market plunges and a Wall Street trader suffers a heart attack when he hears about it, the cardiac event is a translation of a social stress event into physiology. The human brain is very good at that— the translation of psychological input into physiological events. And a good part of our psychological input comes out of the culture we live in or the social circumstances in which we find ourselves. It's a fantasy, for example, to believe that the condition of poverty does not affect the physiology of the people in the midst of it.

The idea that psychological stress can cause disease had its modern heyday in the 1930s. By the 1960s, the idea had been cast aside, mostly with a popular misconception that a "psychosomatic" illness was no more than malingering. Then in the 1980s, with a more sophisticated understanding of biochemistry and molecular biology, researchers came to realize that psychological stress activates the hypothalamic-pituitary-adrenal (HPA) axis, causes secretion of several powerful "stress" hormones, and such hormones under certain circumstances can have adverse effects on organ physiology. Stress research and psychosomatic research are now overlapping domains.[42]

Maternal stress can have many causes, ranging from natural disasters like 9/11 and earthquakes to local family or marital problems. A whole cluster of causes are culture-dependent, particular stress produced by what has been called the "culture of poverty"—psychological stress that derives from living on the edge of economic survival, or living in a crime-ridden neighborhood where even physical survival is uncertain. Humans have physiological mechanisms that determine how the body reacts to psychological stress, and activation of these mechanisms in the mother can seriously affect fetal development.

The major mechanisms involve the HPA axis. The hypothalamus, pituitary, and adrenals that form the HPA axis also collectively form the neuroendocrine control system. The hypothalamus and pituitary gland, adjacent to each other and connected by nerve fibers, are part of the evolutionary "old brain"—a group of anatomic regions also present in lower animals. The adrenal glands (suprarenal glands) are far away from the brain at the top of each kidney.

In the simplest HPA-axis scheme, the hypothalamus receives neural input from other parts of the brain plus chemical input from hormones (chemical messengers) in blood, then secretes its own hormones (for example, oxytocin and antidiuretic hormone) that are carried directly to the nearby pituitary gland. In response, the pituitary gland in turn secretes specific hormones into the blood that affect the adrenal glands. The adrenal glands in turn secrete their own hormones that among other things affect the hypothalamus and pituitary in what engineers call a feedback loop. One of the major hormones secreted by the adrenal glands is cortisol, also called the stress hormone because it's involved in the physiological response to stress. Cortisol acts as a blood messenger to increase blood pressure, blood sugar level, and to suppress certain immune reactions.

The HPA axis is only part of the neuroendocrine system working in the body, but it's a major part and it affects nearly everything fundamental happening in human physiology, including digestion, metabolic rates, sexuality, mood, the immune system, and reactions to physical and psychological stress. Psychological stress has been studied as a cause or exacerbating factor in physical diseases as different as mononucleosis, diabetes, hypertension, coronary heart disease, cancer, and so on.[43]

Stress hormones can also be powerful modifiers of fetal development, which is how the psychological state of a pregnant mother can affect the physiological state of her fetus. Quantitative measurement of postnatal behavioral and IQ consequences are problematic, but there's enough evidence that makes biological sense to underscore the importance of maternal stress as a mediating variable between the maternal psychological and social environment and the development of the fetal body and brain.

In the fetus, the adrenal glands grow fast, are of large size, and are extremely active. They also use the large amounts of progesterone supplied by the placenta to synthesize cortisol, but most of the cortisol present in fetal blood is apparently from the mother's blood—at least in experiments using nonhuman primates.[44]

Cortisol is a powerful steroid hormone that in the fetus affects every physiological system. It also affects the development of the brain. It's likely that the level of cortisol in fetal blood is constrained to a certain range by feedback mechanisms. This implies that when an increase in maternal blood cortisol is reflected in an increase in fetal blood cortisol, the production by the fetus of its own cortisol will be reduced.

The principle is homeostasis—the maintenance of physiological stability by control mechanisms. It's an operating principle prenatally as well postnatally, except that prenatally homeostasis has to provide the conditions required for gestation, conditions for the rapid development of cells and tissues—including the cells and tissues of the developing brain.

Control is the essence here. The development of the fetus is steered by control mechanisms—and the HPA axis is one of those mechanisms.

Cytokines—regulatory proteins released by cells—take part in cell growth and cell differentiation, and they have strong involvement with the HPA axis. In humans, psychosocial stress causes increases in serum levels of cytokines, and when such increases occur in a mother during pregnancy, the result can be a damaging impact on the developing fetus.[45] Indeed, every stress interchange between mother and fetus may influence fetal growth and brain development.[46]

In general, if a mother is stressed while pregnant, her child is substantially more likely to have emotional or cognitive problems, including an increased risk of attention deficit hyperactivity disorder, anxiety, and language delay—and these outcomes are apparently independent of any maternal postnatal depression and anxiety.[47]

Episodes of acute psychosocial maternal stress during pregnancy are of special significance. Stress experiences during pregnancy can affect postnatal levels of neurological development and cognitive performance. Disasters, natural or man-made, provide striking examples. In 1976, China suffered a powerful earthquake that killed an estimated 250,000 people. Twenty-three years after that earthquake, significant differences apparently existed in China between the group in fetal development during the earthquake and a group born 1 year later—differences in intellectual functioning, depression, and the size of certain brain regions.[48] During the Chernobyl radiation disaster, prenatal exposure to maternal stress in the second trimester of pregnancy may have resulted in prenatal programming of physiological systems that control cortisol and testosterone levels in the fetus and in offspring.[49] The effects of natural disasters on fetal development are a signal to us that maternal stress during pregnancy, no matter what the cause, is of consequence to the fetus.

Maternal stress increases the probability that offspring will develop schizophrenia. The Dutch Hunger Winter was one example of this.[50] There are many examples and much speculation.[51] As usual, the main research problem is the long delay between maternal stress during pregnancy and the emergence of psychopathology in adult offspring.

Posttraumatic stress disorder and other affective disorders may both act as maternal environmental challenges affecting the fetus and may themselves be more likely in offspring made vulnerable by maternal stress during fetal development.[52]

In general, culture can shape how people respond to personal stress. At a personal level, any stressful event affecting the mother—such as divorce, death of a husband during pregnancy, or job loss—can also affect fetal development. Such stresses have been associated with increased rates of depression, schizophrenia, and criminality in adult offspring.[53] There's also evidence that maternal-stress impact on fetal development during the third trimester is associated with postnatal symptoms of depression and ADHD.[54] In addition, chronic maternal anxiety causes an increased stillbirth rate, fetal growth retardation, and changes in the placenta.

One important measure of the effects of maternal stress during pregnancy involves infant temperament. Maternal stress during pregnancy predicts cognitive performance deficits and fearfulness in infancy.[55] Measures of temperament and behavioral reactivity in the first few years of postnatal life show the influence of prenatal stress and maternal-placental hormones.[56]

It's possible that fetal overexposure to stress hormones causes physiological programming that endures through childhood and adulthood. Animal experiments demonstrate that prenatal stress hormone excess reduces birth weight and causes lifelong postnatal hypertension, hyperglycemia, and behavioral abnormalities.[57] In fact, the scientific literature cites many animal experiments on the effects of maternal stress on fetal development, but it's difficult to interpret what these results mean for human fetal development.[58]

Nonetheless, the programming principle is clear: if the fetus experiences high cortisol levels, the later production of cortisol in response to stress by the child and adult may be permanently changed. In female offspring that in turn become pregnant, this change may itself impact the next generation of fetuses. Psychosocial stress can have a programming developmental effect on cortisol reactivity.[59] If such programming enhances the physiological response to stress of offspring beyond safe levels, the outcome may be physical disease. Physiological changes in neuroendocrine systems may predispose offspring to cardiovascular disease through influence on plasma glucose, lipid concentrations, and blood pressure. The combination of enhanced stress susceptibility and ordinary psychosocial stressors may be an important component of disease risk in human populations.[60]

In general, prenatal exposure to excess glucocorticoids (for example, cortisol) can program brain function in adult life, predisposing the individual to some pathologies and maybe protecting the individual from other pathologies. Some types of programming may be transmitted to subsequent generations.[61]

The physiology and biochemistry of the human neuroendocrine system is complex, highly variable between individuals, and subject to impacts from the maternal environment. The evidence that the activity of the maternal HPA-axis influences the development of the fetal brain is clear and expected. The fetus does not develop in a bubble that isolates it from the mother: the dynamics of maternal stress physiology, whether internally driven or driven by the outside world, must be a factor in the development of the fetal brain and nervous system.

THE CULTURE OF POVERTY AGAINST THE FETUS

On July 10, 2007, President George W. Bush, in a speech in a hotel in Cleveland, said: "I mean, people have access to health care in America. After all, you just go to an emergency room."[62]

Although this is one of the silliest statements ever made about health care, it may reflect the views of many people who never think about problems of obtaining health care because they can afford it no matter what it costs.

A pregnant woman cannot go to an emergency room for ordinary prenatal care, which is essentially preventive medicine and not emergency medicine. In America, poor women are dependent on government-funded social-welfare programs for access to health care. Those programs are hardly state-of-the-art medicine. The consequence is a relationship between poverty and medical problems during pregnancy and delivery.[63]

But in all ways, poverty is a condition that readily transforms into an inherited disease. The transformation is manmade and will occur in any society in which the condition of poverty means lack of adequate prenatal care during pregnancy and hazardous exposure to neurotoxins in the environment. In such societies, and America is a good example, poor people never receive as much health care as everyone else. It's poor people who live near waste dumps, not the middle and upper classes. Moreover, daily life among the poor is more stressful, fraught with family conflict and violence between parents who may suffer from chronic anxiety or depression, or whose psychiatric dysfunction may be exacerbated by various poverty-related circumstances. The effects of this stress on pregnant mothers and on their unborn fetuses are notable: significant correlations exist between these factors and low-birth-weight infants among mothers in poverty.[64] Moreover, maternal anxiety, depression, and elevated cortisol in late gestation are associated with negative postnatal infant temperament.[65]

The paradigm for the transformation is plain: low socioeconomic status (poverty) results in early age of pregnancy, poor prenatal care, and negative effects on fetal development (through exposure to maternal stress and distress, and to alcohol, tobacco, lead, and other environmental neurotoxins), which result in lower IQ and higher crime, which result in sustained poverty, which starts the cycle again.

Among academics, a debate continues about whether psychiatric disorders are caused by poverty (causation) or whether individuals with those disorders are socially and psychologically maladapted and therefore poor as a consequence (selection).[66] It seems to me this debate is ridiculous, since given the fetal-impact-poverty cycle, both causation and selection must occur.

Anyone who thinks it's so easy to break out of the cycle of poverty needs to look closely at those living in poverty in the inner cities or in rural regions of the country such as Appalachia. Look at their living conditions, and then look for impacts on fetal development. A pregnant woman looking at you from the doorway of a shack in Chicago or Georgia or Texas is in a sisterhood with a pregnant woman looking at you from the doorway of a shack in West Virginia. The women may be black or brown or white, but it's a sisterhood. And their daughters and granddaughters will be in sisterhoods also. So it goes.

Poverty produces its own culture, its own environment, and in most industrialized countries, and countries that are now rapidly industrializing, the culture of poverty and its environment provide ripe ground for the dissemination of dangerous chemicals in air, water, and food—particularly neurotoxins. The neurotoxins discussed in this book are for the most part community neurotoxins. The degree of exposure and the severity of their impact outcome depend as much on socioeconomic circumstance as on individual biology. Impacts of the prenatal environment on the developing brain create permanent changes in brain structure and brain chemistry, and these changes are reflected in postnatal behavior during childhood and throughout life. But postnatal behavior is also shaped by postnatal environment and socioeconomic circumstances. Psychopathologies produced by fetal neurotoxins are influenced by social environment. Toxicity is not a physical property of a toxin—it's a variable dependent on many conditions, among them socioeconomic circumstance.[67]

For example, many homes of urban families in poverty are infested by cockroaches. In New York City, among African-American and Dominican women in northern Manhattan and the South Bronx, 85 percent report that pest control measures are used in the home during pregnancy, mostly for cockroach control. All of these women (100 percent) have detectable levels of three different pesticides in their blood, and 30 percent of these women have detectable levels of eight pesticides in their blood.[68] Umbilical cord samples show that pesticides are readily transferred to the fetus.[69] Prenatal exposure to pesticides is correlated with fetal growth restriction.[70] In America, pesticide use among poor minority women is a continuing problem with cognitive consequences in children that are still hardly tabulated. How much of IQ depression among the offspring of such women is due to pesticide use in the home? We don't know—but it's certainly worth investigating.

In general, what is important is the degree and kind of social stress and the biological substrate that conditions the response of the individual to stress. We do know that an abrupt change in economic conditions can have striking effects. For example, if in a community in poverty a new local large business moves a great number of families out of poverty by employing people previously unemployed, the psychopathology profile of that community will likely change, with some psychiatric disorders becoming less prevalent but some others unchanged in prevalence.[71] Movement of families out of poverty in this community, by reducing psychological stress, has also reduced mental illness.

It's common that children who are poor have higher levels of depression and antisocial behavior.[72] This is not a problem unique to America. In Australia, for example, the more often families experience low income, the higher the rate of child behavior problems at age 5.[73] Comparable correlations are found nearly everywhere. Prenatal impacts may be one cause. Another cause may be the child's postnatal experience of maternal depression produced by poverty. Owing to social constraints, female children are more likely to remain in poverty than male children. Later, these female children become pregnant and the cycle of impacts begins again.

It seems self-evident that psychiatric disorders have social consequences for the individual. One consequence is truncated education.[74] But the degree and nature of the social consequences of psychiatric disorders vary with socioeconomic status: poor children with psychiatric disorders are not subject to the same constraints and consequences as children with psychiatric disorders in middle or upper class families. Poor children are in a different world. Children, both male and female, born into poverty in Northern Ireland, for example, are at a special risk of developmental delays in motor functions and reading ability.[75] Such children are constrained by their deficits to be poor as adults and have their poverty affect the fetal development of their offspring. Thus the cycle begins again in the next generation.

The culture of poverty encourages alcohol and tobacco use during pregnancy to relieve the stress of daily life. Both alcohol and tobacco use during pregnancy are a consequence of the interplay between psychological and social forces. The correlations are clear: women who use tobacco are twice as likely to have a psychiatric disorder than nonusers, and women who use tobacco during pregnancy are even more likely to have a psychiatric disorder.[76] In the general American population, among pregnant women, 22 percent use cigarettes and 12 percent meet the criteria for nicotine dependence. Among pregnant women with cigarette use, 45 percent meet criteria for at least one mental disorder, and among those with nicotine dependence, 57 percent meet criteria for at least one other mental disorder.[77] Given the impact of tobacco use during pregnancy on fetal development, these are the statistics of a troubled society. Too many women either don't know about the dangers or they do know and they don't care.

Alcohol and tobacco use by women in poverty is not unique to America but endemic throughout the Western world. In Germany, for example, tobacco and alcohol use is more prevalent in lower socioeconomic groups and particularly high among the unemployed and among people living alone. In Germany, people in poverty spend up to 20 percent of their income on tobacco.[78]

Tobacco and alcohol use are also correlated with heavy caffeine use (more than three caffeinated drinks a day) among pregnant women before and during pregnancy. Older women are more likely to smoke and ingest caffeine or drink alcohol and ingest caffeine during pregnancy. Caucasian women are more likely to continue smoking during pregnancy, while African-American women are more likely to continue drinking during pregnancy.[79]

Poverty increases the prevalence of infection by reducing immunological defenses. Maternal and fetal infection during pregnancy is common among groups in poverty. The combination of poor maternal nutrition and maternal infections during pregnancy are especially powerful impacts on the development of the fetus. These are consequences of social conditions, a direct connection between society and fetal damage.[80]

The Turkheimer study discussed in chapter 10 pointed out that correlations between heredity or environment and IQ are dependent on socioeconomic class, with environment contributing nearly all the variance for the lower socioeconomic class.[81] The study shows us that the contributions of heredity and environment to variance of IQ scores depend on the socioeconomic class of the population tested. The Flynn effect shows us that distributions of IQ scores are artifacts of test construction and depend on the generation tested. Both effects are of great importance for social policy.

Poverty and Adolescent Pregnancy

Adolescent pregnancy is common in groups in poverty, and we need to ask how this affects the fetal environment. What we know is that adolescents who get pregnant are more likely to start smoking early, to abuse alcohol and other drugs, to have low interest in academic performance, to be children in single-parent families, and to be poor. Adolescent pregnancies are prone to complications such as anemia, hypertension, sexually transmitted diseases, and premature delivery. The consequences of the whole package of risk factors are fetuses impacted by growth restriction and infection and exposure to various neurotoxins.[82]

It seems to me that dealing with the problem of adolescent pregnancy by communicating with pregnant teenagers about the importance of prenatal care is equivalent to offering an Aspirin to bring a fever down. We do need to educate teenage girls about the crucial need for prenatal care, but such education does not attack the causes of teenage pregnancy. In America, teenage pregnancy is a cultural phenomenon associated with poverty, familial chaos, and hopelessness. If there's a public desire to reduce teenage pregnancy, is there enough public focus on getting people out of poverty? We are a dismal people if poverty is considered merely collateral damage in a free-market economy.

The poverty connection is paramount and extends across most American ethnic groups in poverty. But the culture of poverty in America may have a special character. For example, low-income Hispanic teenage girls who are more acculturated into American culture have a greater likelihood of drinking alcohol in and around the time of pregnancy than less acculturated Hispanic girls.[83] What are the social forces that produce these consequences? In southern California, 30 percent of white non-Hispanic, black non-Hispanic, and English-speaking Hispanic women drink during pregnancy compared with only 16 percent of Spanish-speaking Hispanic women.[84] We have little chance of reducing the prevalence of fetal alcohol impacts until we understand the social dynamics of alcohol use during pregnancy.

Ideology is a lazy method of dealing with complex social problems. It's a method that allows us to do hardly more than fool ourselves. I've tried to lay out in this chapter how the impact of culture on fetal development can produce cycles of poverty, cycles of misery, the shackling of large numbers of people by the chains of circumstance. More than 160 years ago, Charles Darwin gave us a single sentence to clarify the issue:[85] "If the misery of the poor be caused not by the laws of nature, but by our institutions, great is our sin."

CRIMINALITY AND THE FETAL ENVIRONMENT

Of the many myths that surround the question of heritability of behavior, those about human criminality are almost immutable, fixed in the media and in the public psyche.

In the 1870s, the Italian criminologist Cesare Lombroso became famous for the idea that criminals were evolutionary regressives and could be recognized by atavistic physical features such as large jaws, high cheekbones, hawk-like noses, fleshy lips, and so on.[86] He published monograph after monograph detailing his observations. His ridiculous ideas were influential for more than 50 years. An irony is that he himself had a hawk-like nose and fleshy lips.

In 1877, the American criminologist Richard L. Dugdale published an exhaustive lineage trace, in which he claimed that a descendant of early Dutch settlers born about 1730 had been the ancestor of a family of criminals he called the Jukes (a pseudonym).[87] Family members included more than 76 convicted criminals, 18 brothel keepers, 120 prostitutes, over 200 relief recipients, and 2 cases of "feeble-mindedness." Dugdale believed that criminal behavior derived from both environment and heredity, but his monograph became famous as evidence for criminality as an inherited trait.

In the 1920s, the American eugenics movement incorporated the nineteenth century ideas of hereditary criminality. Those ideas lingered, and by the middle of the twentieth century, despite firm protests by some American psychologists and social workers, the belief that criminality had a largely hereditary origin had become deeply embedded in American culture. Accounts of the Jukes family could be found in nearly every undergraduate college textbook in psychology as an example of inherited criminal and immoral behavior.

The myth entered popular culture as well. In 1954, William March published a novel called *The Bad Seed*,[88] about a young mother who discovers her 8-year-old daughter is a murderer. The mother also discovers that she herself is the daughter of a serial killer. By this time the 8-year-old has killed three people. The mother decides to poison the child and commit suicide. The mother kills herself but the child survives. The novel was a great commercial and critical success and was nominated for the 1955 National Book Award. This novelistic trumpeting of hereditary criminality also became a successful film, a long-run Broadway play, and in 1985, a successful television play.

In 2002, the New York Times published an account of three generations of the Bogle family, a group with 28 members convicted of various crimes.[89] The Times article noted that "crime often runs in families," but contained no discussion of the actual origins of familial crime.

In 2008, the journal *Medical Hypotheses* published an editorial announcing that genetic evidence exists that criminality is inherited.[90] But despite the claim, no genetic "evidence" was presented, only genetic rhetoric. Completely serious, the editorial states: "Rather surprisingly, despite the Darwinian and the Scientific Revolution, most intellectuals and even many scientists are still reluctant to accept some inescapable social implications of Darwin's theory of evolution."

The so-called genetic origins of criminality are of a kind with the genetic origins of happiness. The ease with which some genetic determinists hold up an extremely complex human behavior pattern and claim a genetic inheritance for it is astonishing. And it gets more astonishing as molecular biology continues to demolish the classical idea of a "gene" as a unitary element of heredity.

Genes are not unitary. There are genes involved in numerous traits, and there are traits connected to numerous genes. Transcription often involves assembly of bits and pieces of DNA that do not constitute a unitary gene before transcription—and the assembly can be controlled by interaction with the local intracellular environment.

There are some rare diseases that evidently derive from mutations that obey Mendelian genetics, but there are no complex human behaviors that have such identified origins—and given the significance of epigenetics for phenotype outcomes, there is little likelihood that simple and exclusive genetic origins will ever be found for any complex human behavior, including criminality.

The biology of our behavior does not involve any set of simple rules. Nor is there any experimental basis for the idea that the biological origins of the behavior of all organisms obey the same principles no matter what level of organismic complexity. The origins of behavior may be qualitatively different for different kinds of organism. The wiggling of a roundworm does not derive from the same biological substrate as the wiggling of a human belly dancer. If we want to understand the origins of wiggling, genetic analysis may be useful in the former case but it's unlikely to be useful in the latter case. That's an elementary constraint to which too many genetic determinists seem oblivious.

Meanwhile, what is needed is an understanding of neurological and psychiatric realities. Every fetal environmental impact discussed in this book that we know produces changes in the developing brain and consequent cognitive and emotional deficits has the potential to be related to childhood and adult violence and criminal behavior.

Fetal alcohol exposure, for example, causes physical, neurological, and psychological impairments that often lead to social maladjustment and subsequent disruptive, violent, and eventually criminal behavior.[91] A common FASD impairment is a language deficit that results in poor social communication and consequent social maladjustment.[92] Low IQ caused by FASD exacerbates any impairment and makes social maladjustment more likely.

Fetal lead exposure, endemic in America and elsewhere on a massive scale for nearly 80 years, is also related to violence in children and criminal behavior in adults.[93] The evidence is as follows:

- Analysis of air-lead levels and crime rates in all counties in the contiguous 48 states in the United States, using data from the Environmental Protection Agency, the Census Bureau, and the Federal Bureau of Investigation, shows that air-lead levels have a direct relation to property and violent crime rates.[94]

- A study of nearly 1000 African-Americans from birth to age 22 shows that lead poisoning is among the strongest variables accounting for criminal behavior in male subjects.[95]
- Blood lead levels in preschool children and subsequent crime rates of these children over several decades are correlated in the United States, the United Kingdom, Canada, France, Australia, Finland, Italy, West Germany, and New Zealand.[96]
- In America, long-term trends in population exposure to gasoline lead are closely consistent with changes in violent crime and unwed pregnancy, and long-term trends in paint and gasoline lead exposure are also strongly associated with trends in murder rates going back to 1900.[97]

Attention deficit hyperactivity disorder (ADHD) produced by various prenatal impacts is another variable highly correlated with childhood violence and adult criminal behavior. ADHD is associated with conduct problems, social maladaptation, and delinquent behavior. In the United States, ADHD boys with conduct problems in school are at increased risk for later criminality.[98] This is true not only in America but also in other industrialized countries, for example, in Germany and Sweden.[99]

Of young adult criminals, 51 percent of males and 44 percent of females have a history of childhood psychopathology.[100]

There is also abundant evidence that maternal smoking during pregnancy is associated with later criminal arrests and psychiatric hospitalization of offspring.[101] Some tobacco effects of maternal origin may be mediated by nicotine, since nicotine receptors are already present in the fetal brain during the first trimester.[102] Other tobacco effects may derive from volatile organics in tobacco smoke. The fetal environment is a filtered maternal environment, and filters vary in what they can filter and how well they do it.

BUT WHAT ABOUT THE MORALITY INSTINCT?

Underlying the myth of innate criminality is another myth about the innateness of social feelings and the moral sense. For example, in Charles Darwin's *The Descent of Man* (1879), his account of human evolution and the evolution of human behavior,[103] he postulated human morality as an evolved (and therefore inherited) trait. As evidence he presented rhetoric rather than science—and page after page of silly anthropomorphism: "It is certain that associated animals have a feeling of love for each other, which is not felt by non-social adult animals."[104] For humans

> As man is a social animal, it is almost certain that he would inherit a tendency to be faithful to his comrades and obedient to the leader of his tribe, for these qualities are common to most social animals.[105]

So began the modern idea of a moral instinct and hereditary rectitude.

And now it's claimed that there is increasing scientific evidence to support this notion. The reality is otherwise. What is increasing is rhetoric, not evidence. Most of the rhetoric is in the media and from a small band of evolutionary psychologists and sociobiologists who emphasize evolved inherited hard-wiring as the basis for human behavior. The typical procedure is to hypothesize some particular behavior as an evolutionary adaptation of prehistoric humans, and then conclude that empirical existence of the behavior as a universal or quasi-universal "tendency" in modern humans confirms the starting hypothesis and implies important current social realities—including the hard-wiring of many behaviors by genetic evolution: romance, aggression, rape, altruism, morality, gambling, political views, the keeping of animal pets, and so on.

For example, in one study, an observer watches as preverbal infants differentiate which wooden blocks ("individuals") help other blocks and which wooden blocks hinder other blocks in a "climbing" effort up an incline. The observer notes which blocks the infants look at and how they look at them and concludes that the infants find the helper blocks more appealing.[106] From this the observer further concludes that infants engage in social evaluation, and that evaluating individuals on the basis of their social interactions is universal and unlearned. And the grand conclusion is that "this capacity may serve as the foundation for moral thought and action, and its early developmental emergence supports the view that social evaluation is a biological adaptation."[107]

The experiments in this example involved a small group of infants (no more than 16) sitting on the laps of their mothers, with "fussy" infants excluded from the experiments. But more important than the problematic design of these experiments are their conclusions, because it's such conclusions that are picked up by people in other fields and by the media, and sooner or later there's a press report that a "universal moral instinct" as an evolutionary biological adaptation is already evident in infants 10 months old.

This is not science, it's pseudoscience. In the last century, the same experimental observations might have been interpreted by Freudian psychologists in a totally different way and with no more justification. In the science of human behavior, there has always been too much servility to our pet preconceptions. A generation passes, the preconceptions collapse, and the servility starts again from a new platform. There's much in evolutionary psychology that's not pseudoscience, but unfortunately theres enough to be worrisome.

My favorite anecdote about moral instincts involves cats. When a pregnant female cat gives birth to her kittens, she immediately proceeds to lick away the thin membranes that coat each newborn kitten. If she doesn't do this, the kittens will quickly die. For centuries, this behavior of female cats and other female animals was hailed as a sterling example of a "maternal instinct" and also a "moral instinct," since what else could the mother cat's behavior be except a nurturing instinct for her kittens? But physiologists are hard-bitten realists, and as physiological science flowered in the late nineteenth and early twentieth centuries, it wasn't long before someone discovered that a pregnant cat has a physiological salt deficiency, and that the reason the mother cat licks the membranes away from her newborn kittens is that the membranes have a high salt content. In fact, if you soak a tennis ball in a salt solution and present it to the mother cat, she will eagerly lick the tennis ball and allow her kittens to die.

So much for maternal or moral instincts. The salt deficiency in a pregnant female cat most certainly involves a complex of evolved genes and gene regulators. But what has evolved is not the behavior but the physiology, maybe an evolved regulation of the pregnant cat's kidney function. Genes and gene regulators shape only physiology and biochemistry. What we call "behavior" is an outcome of an interaction of biology and environment. Behavior is not a process genetically produced in isolation from environment. What the female cat inherits is a salt deprivation during pregnancy. How we describe her consequent behavior is our problem and not nature's problem. We can call it anything we like, maternal instinct, moral instinct, licking instinct, or whatever—it's physiological salt deprivation that's producing the behavior, not any instinct label that we stick on it.

Too often our reluctance to peel away the labels we use is hazardous to ourselves and to society at large. An example of the dangers of media and public confusion about "instincts" is the insidious shaping of public attitudes towards criminality. If it's believed that a "moral instinct" is already hard-wired at birth, and it's also believed that hard-wiring of such social attitudes is determined by genes, then the implication for the public is that the moral instinct or moral sense or the trait of "morality" is hereditary. This idea further implies that immoral behavior is hereditary, that it will run in families, and that such behavior can be modified only with difficulty if at all. According to this view, public money spent to reduce crime by improving environment is therefore wasted, and in general so is any money spent for rehabilitation. Maybe, according to this view, the unspoken final solution to criminality is eugenics redux, the sterilization of criminals.

Such is the pattern these days for how science (or pseudoscience) segues into politics and public policy.

CODA

What I have tried to elaborate in this book is the reality that no fetus develops in isolation from the community around it.

The most important consequence of this reality is that every fetus is exposed to environmental effects that have the potential to shape, divert, or derange its development. And that potential can have far-reaching effects on that child's brain and behavior in postnatal life. The classical idea of "immaculate gestation" thus needs to be abandoned if we're to have a full understanding of the origins of human intelligence and behavior.

A variety of people, including many in industry, may be opposed to the idea of the fetus as an environmental target.

Others may be opposed to the idea that childhood and adult behavior can be dramatically shaped by what happens in the prenatal environment. But I do think the facts and numbers reviewed in this book tell us otherwise.

As for academic controversy, it always lurks in the background. For genetic-determinist conservatives, the only IQ and behavior determinant believed important has always been genetics—with some genetic determinists even promoting the silly extreme view of innate white supremacy. It's a sorry history. I think too many genetic determinists have gone beyond reason and science in their dictum that genes rule. Their drumbeat is to ignore all environment, both prenatal and postnatal.

But some researchers who emphasize postnatal environmental determinants of behavior have also had a mistaken drumbeat. Thirty-five years ago, some progressive psychologists dismissed the prenatal environment as an important source of group differences in IQ.[108] They believed any focus on the prenatal environment was a genetic-determinist conservative diversion that would reduce the need for policies to address group social and economic differences. During the years, they have overlooked the fact that for most disadvantaged groups the prenatal environment is essentially controlled by social, political, and economic forces. The prenatal environment routinely translates social and economic conditions into uterine biological variables that can produce transgenerational cognitive dysfunctions and deficits. Attending to those conditions adequately would benefit everyone, whether a fetus, a pregnant mother living in dire poverty, or an entire overburdened community.

In a similar way, some religionists have also sounded a mistaken drumbeat. A great deal of emotional and political energy in America is exhibited by certain religious groups in their promoted concern for the life of the fetus—the life of the unborn. It's a disappointment that so little of that energy is devoted to rectifying man-made social and industrial conditions that mangle the unborn in America by the millions.

What seems certain to me is that accepting things as they are is a defeat of the human spirit. But I don't think such defeat is necessary. I'm optimistic about what mankind can and will do to eliminate preventable fetal impacts that damage the brains and bodies of generations of children. It's a tractable problem, much of it man-made.

NOTES

CHAPTER ONE: THE RICHNESS OF OUR IGNORANCE

1. Berkowitz et al., 2003.

2. Engel et al., 2005; Wolff et al., 2005; Yehuda et al., 2005.

3. Levine et al., 2007; Sarkar et al., 2007, 2008.

4. Gluckman et al., 2005.

5. Darwin, 1859, p. 459.

6. Bryan, 1924.

7. Wilson, 1975; Dawkins, 1976.

8. Wilson & Wilson, 2007.

9. For a summary of some new ideas about levels of evolution, see Jablonka & Lamb, 2005, 2007.

10. Couzin, 2002.

11. Turkheimer et al., 2003.

12. Grigg, 2004.

13. Hampton, 2004.

14. Opler et al., 2005.

15. A "risk factor" is a measure of increased risk when comparing, for example, the risk of an exposed group to a nonexposed group. A risk factor of 2 means the exposed group has twice the likelihood of an observed effect of exposure than the nonexposed group.

16. Patterson, 2007.

17. Grandjean et al., 2008.

18. Balter, 2008; Fodor, 1998.

19. Sadler, 2006, p. 111.

20. Classical ideas about instincts can be found in Beach, 1951; see also Mameli & Bateson, 2006.

21. Agin, 2008b; Harmon, 2007; Kaplan & Rogers, 2003; Nelkin & Lindee, 2004.

CHAPTER TWO: POLLUTION BABIES

1. Lidsky & Schneider, 2005; also described in Agin, 2007.

2. Levin et al., 2008.

3. Canfield et al., 2003; Wang et al., 2008.

4. Lanphear et al., 2005.

5. EPA, 2007.

6. EPA, 2007.

7. Zierold & Anderson, 2004.

8. CDC, 2005.

9. Whitehead & Leiker, 2007.

10. CDC, 2007.

11. Blake-Gumbs, 2006.

12. American Academy of Pediatrics, 2005.

13. Hu et al., 2006.

14. Tong et al., 2000.

15. Mielke, 1999.

16. Lidsky & Schneider, 2003.

17. The two modern signpost arguments of hereditarian psychometrics concerning IQ differences between groups are Jensen, 1969 and Herrnstein & Murray, 1994.

18. Mielke, 1999.

19. Laraque & Trasande, 2005.

20. Chen et al., 2007.

21. Yuan et al., 2006.

22. Needleman et al., 1996.

23. Opler et al., 2004.

24. Bellinger, 2004.

25. Agin, 2008c.

26. Grandjean & Landrigan, 2006.

27. Khattak et al., 1999.

CHAPTER THREE: FROM ONE CELL TO A HUNDRED TRILLION

1. Steptoe & Edwards, 1978.

2. Conley et al., 2006.

3. Amaral et al., 2008.

4. Pennisi, 2007.

5. Judson, 2001.

6. Arndt, 2007.

7. Bird, 2007.

8. McGowan et al., 2008.

9. Certain cells of the immune system have a modified genome and are exceptions to the rule of identical genomes in cells of various types. Also, each ordinary cell has two copies of the genome, and the copies usually have slight differences.

10. Yelin et al., 2005, 2007.

11. Gilbert, 2006.

12. For an account of Haeckel's writings on eugenics, see Kaplan and Rogers, 1994.

13. Goodman & Coughlin, 2000; Honeycutt, 2008; Koentges, 2008.

CHAPTER FOUR: THE FETAL BRAIN

1. Kenneally, 2006.

2. Devlin et al., 2003; Korkman et al., 2005; Pulsifer et al., 2004.

3. Gould et al., 1999.

4. Cameron & Dayer, 2008.

5. Harvey & Svoboda, 2007; Sabatini, 2007.

6. Glazier et al., 2002.

7. Martin, 2007.

8. Purves, 2007; Sherrington, 1937.

9. Grove, 2008; Mangale et al., 2008.

10. Robinson, 2005.

11. Johnson, 2005.

12. Nadarajah et al., 2003.

13. Samuels & Tsai, 2004.

14. Rakic & Lombroso, 1998.

15. Edelman, 1987.

16. Stevens et al., 2007.

17. Ruediger & Bolz, 2007.

18. Whittle et al., 2007.

19. Rakic & Lombroso, 1998.

20. Kerjan & Gleeson, 2007.

CHAPTER FIVE: LIFE IN UTERO: SHAPING OR DESTRUCTION?

1. Van der Zee, 1982.

2. Lumey & Van Poppel, 1994.

3. Barker, 1981; Barker et al., 1982; Barker & Osmond, 1987; Bateson et al, 2004.

4. Barker, 1981.

5. Landrigan et al., 2005.

6. Nafee et al., 2008; Reik et al., 2001.

7. Hanson & Gluckman, 2008.

8. Some behavioral geneticists are beginning to understand that analysis of the impacts of environment on gene expression may be as important as the "identification" of genes associated with behavior. See Rutter, 2006, pp. 211–220.

9. Sadler, 2006, p. 40.

10. Rees & Inder, 2005.

11. Nathanielsz, 2000.

12. Hong et al., 2005.

13. Gottesman & Gould, 2003.

14. Rees et al., 2008.

15. Rees & Harding, 2004.

16. Harper, 2005.

17. Tustin et al., 2004.

18. Muñoz-Tuduri & Garcia-Moro, 2008.

19. Brunton & Russell, 2008.

20. Jansson & Powell, 2007; Ventolini & Neiger, 2006.

21. Innis, 2007.

22. Rees & Inder, 2005.

23. Sadler, 2006.

24. Rees et al., 2008.

25. Rees & Inder, 2005.

26. Toro et al., 2008.

27. Richardson et al., 2008.

28. Richardson et al., 2008.

29. Schuetze et al., 2007.

30. Ministry of the Environment, 2002; Ui, 1992.

31. Ui, 1992.

32. Nierenberg et al., 1998.

33. Grandjean, 2008.

34. Choi, 1988.

35. Choi, 1988.

36. Griffiths et al., 2007; Trasande et al., 2006, 2007.

37. Trasande et al., 2005, 2006.

38. Trasande et al., 2006.

39. Axelrad et al., 2007.

40. Grandjean & Landrigan, 2006.

41. Claudio, 2002.

42. Jacobson & Jacobson, 1996.

43. Faroon et al., 2001.

44. Gray et al., 2005; Stewart et al., 2005, 2008.

45. Borrell et al., 2004.

46. McLachlan et al., 2006.

47. Anway et al., 2005.

48. Ritz & Wilhelm, 2008.

49. Perera et al., 2002.

50. Wang & Pinkerton, 2007.

CHAPTER SIX: THE ENDLESS FETAL HANGOVER

1. Hardman et al., 1996, p. 386.

2. Rorabaugh, 2003.

3. Dudley, 2004.

4. Gould & Lewontin, 1979.

5. May & Gossage, 2001.

6. Fox & Druschel, 2003.

7. Wattendorf & Muenke, 2005.

8. Jones et al., 1973.

9. Jones & Smith, 1973.

10. Jones & Smith, 1973.

11. Sood et al., 2001.

12. Olney, 2004.

13. Wattendorf & Muenke, 2005.

14. Nathanielsz, 1999, p. 331.

15. United States Department of Health and Human Services, 2005.

16. Bobo et al., 2006; CDC, 2004d; Chambers et al., 2005.

17. Floyd et al., 1999.

18. O'Connor & Whaley, 2003.

19. Chambers et al., 2005.

20. Tsai et al., 2007, 2007b.

21. Ebrahim et al., 2000.

22. Sampson et al., 1997.

23. CDC, 2002.

24. Willford et al., 2006.

25. In this context, "set" refers to a readiness to think or perceive in a certain way.

26. Rasmussen, 2005.

27. Watson & Wesby, 2003.

28. Bailey et al., 2004.

29. Streissguth et al., 1990.

30. Testa et al., 2003.

31. Schonfeld et al., 2005.

32. Sokol, et al., 2003.

33. Molina et al., 2007.

34. Bookstein et al., 2002.

35. Mattson et al., 1992.

36. Cortese et al., 2006.

37. Cuzon et al., 2008.

38. Jacobson et al., 2008.

39. Kodituwakku, 2007.

CHAPTER SEVEN: UNBORN DAYS AND SEXUALITY

1. Michel et al., 2001.

2. Segal, 2006.

3. Galen, 1968; Gilbert, 2006, p. 773.

4. Gilbert, 2006, p. 774.

5. Wilson & Davies, 2007.

6. Hines, 2006. For an accessible review of genes, hormones, and the science of sex differences, see Rogers, 2001.

7. Zhou et al., 1995.

8. Money, 1975.

9. CBC News, 2004.

10. Bradley et al., 1998.

11. Segal, 2006.

12. Cohen-Bendahan et al., 2005; Dörner et al., 2001.

13. Gooren, 2006.

14. Rahman, 2005.

15. James, 2005.

16. McConaghy & Blaszczynski, 1980.

17. Hershberger & Segal, 2004.

18. Hall & Love, 2003.

19. Berenbaum & Hines, 1992.

20. Rahman, 2005.

CHAPTER EIGHT: DEVELOPMENTAL BRAIN DISABILITIES

1. Boyle et al., 1994; Skounti et al., 2007; Yeargin-Allsopp et al., 2003.

2. Fornoff, et al., 2007.

3. Michigan Department of Community Health, 2008.

4. CDC, 2008c.

5. Fornoff et al., 2007.

6. Needleman et al., 1979.

7. CDC, 2004.

8. Silvert, 2000.

9. Rosen & Hobel, 1986.

10. Paneth et al., 2005.

11. An exception to this would be a rare recessive genetic condition.

12. Rezaie & Dean, 2002; Yoon et al., 2003.

13. Temkin, 1945.

14. CDC, 2008.

15. Scher, 2003.

16. Procopio & Marriott, 1998.

17. Procopio et al., 1997.

18. Sun et al., 2008.

19. Suzuki et al., 2008.

20. Brooks et al., 2007, p. 442.

21. Briellmann et al., 2001.

22. Mervis et al., 2000.

23. Boyle et al., 1996.

24. CDC, 2004b.

25. CDC, 2004c.

26. Colugnati et al., 2007.

27. U.S. Preventive Services Task Force, 2006.

28. Coggins et al., 2007.

29. Lewis et al., 2007.

30. Bhate & Wilkinson, 2006.

31. Paulesu et al., 2001b.

32. Olson, 2006.

33. Shaywitz et al., 2008.

34. Habib, 2000.

35. Paulesu et al., 2001.

36. Siok et al., 2008. Such thinning of the cortex apparently involves

a pruning of circuits, the pruning increasing analytic efficiency. In general, it is possible to show that in certain kinds of circuits less is more.

37. Jacobson & Jacobson, 1996; Schettler, 2001; Vreugdenhil et al., 2002.

38. Olson, 2006.

39. Hardell et al., 2002.

40. Laucht & Schmidt, 2004.

41. Braun et al., 2006.

42. Nigg et al., 2008; Wang et al., 2008.

43. Braun et al., 2006.

44. Braun & Lanphear, 2007.

45. Banerjee et al., 2007; Shaw et al., 2007.

46. Millichap, 2008.

47. Quoted by Metacritic, 2008.

48. Bakare & Ikegwuonu, 2008.

49. Kanner, 1943.

50. Yeargin-Allsopp et al., 2003.

51. CDC MMWR, 2007b.

52. Pollak, 1997, pp. 249–285.

53. Pollak, 1997, p. 411.

54. ABC News, 2007.

55. Ritvo et al., 1985.

56. Prescott et al., 1999.

57. Sadock & Sadock, 2003, p. 1209.

58. Eichler & Zimmerman, 2008; Muhle et al., 2004.

59. Roohi et al., 2009.

60. London & Etzel, 2000.

61. For the clinical limitations of behavioral genetics, see Tobin, 1999.

62. Minshew & Williams, 2007.

63. Bonilha et al., 2008.

64. Buccino & Amore, 2008; Iacoboni & Dapretto, 2006.

65. Geschwind & Levitt, 2007.

66. Gillberg, 1999; Gillberg et al., 1991.

67. Previc, 2007.

68. Schanen, 2006.

69. Zhao et al., 2007.

70. Croen et al., 2007.

71. Daniels et al., 2008.

72. DeSoto & Hitlan, 2007.

73. Schechter & Grether, 2008.

74. Hertz-Picciotto et al., 2008.

75. Roberts et al., 2007.

76. Sullivan, 2008.

77. Caldwell et al., 2005.

78. Lidsky & Schneider, 2005.

79. Bransfield et al.,2008.

80. Bouchard & McGue, 2003.

81. Altevogt et al., 2008.

CHAPTER NINE: GENES, THE WOMB, AND MENTAL ILLNESS

1. Kessler & Wang, 2008.

2. Canino & Alegria, 2008.

3. Craddock & Owen, 2007.

4. Freedman et al., 1976, p. 418.

5. Szasz, 1973, p. 113.

6. Sadock & Sadock, 2003, p. 472.

7. Sadock & Sadock, 2003, p. 536.

8. Sadock & Sadock, 2003, p. 535.

9. Sadock & Sadock, 2003, p. 508.

10. Arieti, 1955. p. 9.

11. Craddock et al., 2006.

12. Patterson, 2007.

13. Palomo et al., 2004.

14. Lewontin, 2000, p. 97.

15. Rapoport et al.,2004.

16. Howes et al., 2004.

17. Harrison & Weinberger, 2005.

18. Gottesman, 1991, p. 110; Tsuang, 2000.

19. Patterson, 2007.

20. Tsuang, 2000.

21. Sullivan et al., 2003.

22. Sullivan et al., 2003.

23. Petronis, 2003; Petronis, 2004; Petronis et al., 2003.

24. Singh et al., 2002.

25. Robert, 2000.

26. Torrey et al., 1997.

27. Brown et al., 1995; Susser et al., 1996.

28. Zammit et al., 2007.

29. Watson & McDonald, 2007.

30. Gallagher et al., 1999.

31. Opler & Susser, 2005.

32. Mednick et al., 1988.

33. Edwards, 2007.

34. Urakubo et al., 2001.

35. Cannon et al., 2002; Van Erp et al., 2002.

36. Schmidt-Kastner et al., 2006.

37. Boog, 2004; Van Os & Selten, 1998.

38. Khashan et al., 2008.

39. Opler & Susser, 2005; Opler et al., 2004.

40. Brown, 2008.

41. Marx & Lieberman, 1998; Stevens, 2002.

42. Wei & Hemmings, 2005.

43. Dohrenwend et al., 1992.

44. Kajantie, 2006; Seckl & Meaney, 2006.

45. Raison & Miller, 2003.

46. De Weerth et al., 2003.

47. Talge et al., 2007.

48. Pesonen et al., 2006.

49. Rice et al., 2007.

50. Werner et al., 2007.

51. DiPietro et al., 2008.

52. Werner et al., 2007.

53. Wachs et al., 2005.

54. Caspi & Silva, 1995.

55. Belmaker & Agam, 2008.

56. Gale & Martyn, 2004.

57. Bellingham-Young & Adamson-Macedo, 2003.

58. Patton et al., 2004.

59. Andreasen & Black, 2001, p. 515.

60. Bulik et al., 2006.

61. Sadock & Sadock, 2003, p. 739.

62. Kinder, 1997.

63. Rezaul et al., 1996.

64. Castrogiovanni et al., 1998.

65. Favaro et al., 2006.

66. Montgomery et al., 2005.

67. Procopio & Marriott, 2007.

CHAPTER TEN: MUCH ADO ABOUT IQ

1. American Psychological Association, 1996.

2. Platt & Bach, 1997; Richardson, 1998, p. 160.

3. Bouchard & McGue, 2003.

4. I have written at length about this elsewhere, in Agin, 2006, p. 258.

5. Kamin, 1974; Lewontin et al., 1984, p. 101.

6. Lerner, 2006.

7. Lewontin, 1982, p. 71.

8. Downes, 2004.

9. Kempthorne, 1997; Lewontin, 1974.

10. The constraints here arise from elementary mathematical analysis. Given a function $F(x,y)$, without knowing the actual function one can assume a series polynomial expansion of the function. If the series rapidly converges, there might be some basis for using a first order approximation, which involves the first two linear terms in the variables, while discarding the remaining nonlinear terms. The method can be harmless as a guide for future research—but extremely dangerous as a guide for public policy. The problem is that if the series does not rapidly converge you can be in deep trouble. Also, see Layzer, 1974.

11. Haier et al., 2004.

12. Schoenemann et al., 2000.

13. Agin, 2006, p. 258.

14. Shaw, 2007.

15. Grigg, 2004.

16. Farber, 1981, pp. 167–212; Joseph, 2001, 2002; Richardson & Norgate, 2005.

17. CDC, 2008b.

18. Ronalds et al.,2005.

19. Machin, 1996.

20. Schoenwolf et al., 2009, p. 181.

21. Munsinger, 1977.

22. Sadler, 2006, p. 106.

23. Skipper, 2008.

24. Bouchard, et al., 1990.

25. Sulloway, 2007.

26. Jacobs et al., 2001.

27. Richardson & Norgate, 2005. 2006.

28. Devlin et al., 1997.

29. Herrnstein & Murray, 1994, p. 108.

30. Lewontin, 1974.

31. Mandler, 2001.

32. This IQ thought experiment first appeared in a column by the author in *The Huffington Post*, January, 28, 2008. See Agin, 2008. A related analysis can be found in Richardson, 1998, p. 153.

33. See Richardson & Norgate, 2006, for a review of the problems of adoption studies in determining genetic origins of behavior and IQ.

34. Agin, 2007b; Turkheimer et al., 2003.

35. Flynn, 1984.

36. Flynn, 1987.

37. Flynn, 2007.

38. Daley et al., 2003.

39. Meisenberg et al., 2005.

40. Dickens & Flynn, 2001; Flynn, 2007; Neisser, 1998.
41. Flynn, 1998.
42. Russell, 2007.

CHAPTER ELEVEN: CULTURE, POVERTY, AND FETAL DESTRUCTION

1. Ceccanti et al., 2007.
2. May et al., 2006.
3. Colman, 2006, p. 184.
4. Windham & Fenster, 2008.
5. Avila & Blumberg, 2008.
6. WHO, 2007b.
7. CDC, 2002c.
8. Alexander et al., 2008.
9. Kramer et al., 2006.
10. Kierans et al., 2008.
11. CDC/NVSS, 2007.
12. CDC/NCHS, 2008; Gould et al., 2003.
13. Srám et al., 2005.
14. Bobak, 2000; Schell et al., 2006.
15. Wang & Pinkerton, 2007.
16. McDaniel, 2006.
17. CDC/NCHS, 2008.
18. WHO, 2007a.
19. CDC, 2002b.
20. Singh & Kposowa, 1994.

21. Bove et al., 2002.

22. Swan et al., 1998; Waller et al., 1998.

23. Siddiqui et al., 2008.

24. Kenneson & Cannon, 2007.

25. Trabert & Misra, 2007.

26. Gortmaker, 1979.

27. Savage et al., 2007.

28. Alderliesten et al., 2007.

29. Huang et al., 2008.

30. CDC/NCHS, 2008.

31. Aagaard-Tillery et al., 2008.

32. Jauniaux & Burton, 2007.

33. Habek, 2007.

34. Cowperthwaite et al., 2007.

35. Habek et al., 2002.

36. Detjen et al., 2007.

37. Cornelius et al., 2007.

38. Savitz et al., 2008.

39. Bech et al., 2005.

40. Felton et al., 2007.

41. Tamura & Picciano, 2006.

42. Rabkin & Struening, 1976.

43. Istvan, 1986.

44. Mastorakos & Ilias, 2003.

45. Coussons-Read et al., 2007.

46. Correia & Linhares, 2007; Relier, 2001.

47. Meaney, 2001; Talge et al., 2007.

48. King et al., 2000; Watson et al., 1999.

49. Huizink et al., 2008.

50. Fumagalli et al., 2007.

51. Watson et al., 1999.

52. Seckl, 2008.

53. King et al., 2000.

54. Huizink et al., 2004, 2007.

55. Bergman et al., 2007.

56. Wadhwa, 2005.

57. Drake et al., 2007.

58. Merlot et al., 2008.

59. Ouellet-Morin et al., 2008.

60. Phillips & Jones, 2006.

61. Seckl & Meaney, 2006.

62. The White House, 2007.

63. McBarnette, 1987.

64. Astone et al., 2007; Rosen et al., 2007.

65. Davis et al., 2007; O'Connor et al., 2007.

66. Dohrenwend et al., 1992.

67. Weiss & Bellinger, 2006.

68. Whyatt et al., 2002.

69. Whyatt, et al., 2003.

70. Perera et al., 2005; Whyatt et al., 2004.

71. Costello et al., 2003.

72. Mcleod & Shanahan, 1996.

73. Bor et al., 1997.

74. Kessler et al., 1995.

75. McPhillips & Jordan-Black, 2007.

76. Flick et al., 2006.

77. Goodwin et al., 2007.

78. Haustein, 2006.

79. Stotts et al., 2003.

80. Tomkins et al., 1994.

81. Turkheimer et al., 2003.
82. Malamitsi-Puchner & Boutsikou, 2006.
83. Chambers et al., 2005.
84. O'Connor & Whaley, 2003.
85. Darwin, 1845, p. 427.
86. Gould, 1996, p. 152.
87. Dugdale, 1877.
88. March, 1954.
89. Butterfield, 2002.
90. Baschetti, 2008.
91. Fast & Conry, 2004.
92. Coggins et al., 2007.
93. Hwang, 2007; Needleman et al., 1996; Wright et al., 2008.
94. Stretesky & Lynch, 2004.
95. Denno, 1990, p. 84, 106.
96. Nevin, 2007.
97. Nevin, 2000.
98. Satterfield et al., 2007.
99. Dalteg et al., 1998; Retz et al., 2004.
100. Copeland et al., 2007.
101. Brennan et al., 2002.
102. Hellström-Lindahl & Nordberg, 2002.
103. Darwin, 1879.
104. Darwin, 1879, p. 125.
105. Darwin, 1879, p. 132.
106. Hamlin et al., 2007.
107. Hamlin et al., 2007.
108. Kamin, 1974, p. 161.

REFERENCES

Aagaard-Tillery, K. M., Porter, T. F., Lane, R. H., Varner, M. W., Lacoursiere, D. Y. (2008). In utero tobacco exposure is associated with modified effects of maternal factors on fetal growth. Am. J. Obstet. Gynecol. 198:66.e1–66.e6.

ABC News. (2007). Autism times three. December 27. http://abcnews. go.com/GMA/TurningPoints/story?id=4056854&page=1

Agin, D. (2006). *Junk Science: How Politicians, Corporations, and Other Hucksters Betray Us*. New York: Thomas Dunne/St. Martin's Press.

———. (2007). Autism and our passion for simple answers and quick fixes. Huffington Post, June 18, 2007. http://www.huffington-post.com/dan-agin/autism-and-our-passion-fo_b_52651.html

———. (2007b). Genes and IQ: an unsettled Omelet. Huffington Post, October 9, 2007. http://www.huffingtonpost.com/dan-agin/genes-and-iq-an-unsettle_b_67764.html

Agin, D. (2008). Myths of the crazy ape #1: Twins, genes, and IQ. Huffington Post. January 28, 2008. http://www.huffingtonpost.com/dan-agin/myths-of-the-crazy-ape-1_b_83629.html

———. (2008b). Gene-Mongering in the New York Times: How to Twist Science to Suit Your Fancy. Huffington Post, May 27, 2008. http://www.huffingtonpost.com/dan-agin/gene-mongering-in-the-emn_b_103751.html

———. (2008c). Lead redux: an unfinished horror story. Huffington Post, June 4, 2008. http://www.huffingtonpost.com/dan-agin/lead-redux-an-unfinished_b_105249.html

Alderliesten, M. E., Vrijkotte, T. G., Van der Wal, M. F., Bonsel, G. J. (2007). Late start of antenatal care among ethnic minorities in a large cohort of pregnant women. Br. J. Obstetr. Gynecol. 114:1232–1239.

Alexander, G. R., Wingate, M. S., Bader, D., Kogan, M. D. (2008). The increasing racial disparity in infant mortality rates: composition and contributors to recent US trends. Am. J. Obstet. Gynecol. 198:51.e1–51.e9.

Altevogt, B. M., Hanson, S. L., Leshner, A. I. (2008). Autism and the environment: challenges and opportunities for research. Pediatrics. 121:1225–1229.

Amaral, P. P., Dinger, M. E., Mercer, T. R., Mattick, J. S. (2008). The eukaryotic genome as an RNA machine. Science. 319:1787–1789.

American Academy of Pediatrics. (2005). Policy statement. Lead exposure in children: prevention, detection, and management. Pediatrics. 116:1036–1046.

American Psychological Association. (1996). Intelligence: knowns and unknowns. Am. Psychologist. 51:77–101.

Andreasen, N. C. & Black, D. W. (2001). Introductory Textbook of Psychiatry. Washington, DC: American Psychiatric Publishing.

Anway, M. D., Cupp, A. S., Uzumcu, M., Skinner, M. K. (2005). Epigenetic transgenerational actions of endocrine disruptors and male fertility. Science. 308:1466–1469.

Arieti, S. (1955). *Interpretation of Schizophrenia*. New York: Robert Brunner.

Arndt, K. M. (2007). Molecular biology: genome under surveillance. Nature. 450:959–960.

Astone, N. M., Misra, D., Lynch, C. (2007). The effect of maternal socio-economic status throughout the lifespan on infant birthweight. Paediatr. Perinat. Epidemiol. 21:310–318.

Avila, R. M. & Blumber, S. J. (2008). Chronic developmental conditions among Hispanic children in the United States. Health E-Stats. National Center for Health Statistics. March 2008. http://www.cdc.gov/nchs/products/pubs/pubd/hestats/chroniccondi-tions.htm

Axelrad, D. A., Bellinger, D. C., Ryan, L. M., Woodruff, T. J. (2007). Dose-response relationship of prenatal mercury exposure and IQ: an integrative analysis of epidemiologic data. Environ. Health Perspect. 115:609–615.

Bailey B. N., Delaney-Black, V., Covington, C. Y., Ager, J., Janisse, J., Hannigan, J. H., Sokol, R. J. (2004). Prenatal exposure to binge drinking and cognitive and behavioral outcomes at age 7 years. Am. J. Obstetr. Gynecol. 2004 191:1037–1043.

Bakare M. O. & Ikegwuonu, N. N. (2008). Childhood autism in a 13 year old boy with oculocutaneous albinism: a case report. J. Med. Case Reports. 2:56.

Balter, M. (2008). Why we're different: probing the gap between apes and humans. Science. 319:404–405.

Banerjee, T. D., Middleton, F., Faraone, S. V. (2007). Environmental risk factors for attention-deficit hyperactivity disorder. Acta Paediatr. 96:1269–1274.

Barker, D. J. (1981). Geographical variations in disease in Britain. Br. Med. J. (Clin Res Ed). 283;398–400.

Barker D. J., Gardner, M. J., Power, C. (1982). Incidence of diabetes amongst people aged 18–50 years in nine British towns: a collaborative study. Diabetologia. 22:421–425.

Barker D. J. & Osmond, C. (1987). Inequalities in health in Britain: specific explanations in three Lancashire towns. Br. Med. J. (Clin Res Ed). 294:749–752.

Baschetti, R. (2008). Genetic evidence that Darwin was right about criminality: nature, not nurture. Med. Hypoth. 70:1092–1102.

Bateson, P., Barker, D., Clutton-Brock, T., Deb, D., D'Udine, B., Foley, R. A., Gluckman, P., Godfrey, K., Kirkwood, T., Lahr, M. M., McNamara, J., Metcalfe, N. B., Monaghan, P., Spencer, H. G., Sultan, S. E. (2004). Developmental plasticity and human health. Nature. 430:419–421.

Beach, F. (1951). Instinctive behavior: reproductive activities. In: Stevens, S. S. (ed.) *Handbook of Experimental Psychology*. New York: John Wiley & Sons. p. 427.

Bech, B. H., Nohr, E. A., Vaeth, M., Henriksen, T. B., Olsen, J. (2005). Coffee and fetal death: a cohort study with prospective data. Am. J. Epidemiol. 162:983–990.

Bellinger, D. C. (2004). Lead. Pediatrics. 113:1016–1022.

Bellingham-Young, D. A. & Adamson-Macedo, E. N. (2003). Foetal origins theory: links with adult depression and general self-efficacy. Neuro. Endocrinol. Lett. 24:412–416.

Belmaker, R. H. & Agam, G. (2008). Major depressive disorder. N. Engl. J. Med. 358:55–68.

Berenbaum, S. A. & Hines, M. (1992). Early androgens are related to childhood sex-typed toy preferences. Psychol. Sci. 3:203–206.

Bergman, K., Sarkar, P., O'Connor, T. G., Modi, N., Glover, V. (2007). Maternal stress during pregnancy predicts cognitive ability and fearfulness in infancy. J. Am. Acad. Child Adolesc. Psychiatry. 46:1454–1463.

Berkowitz, G. S., Wolff, M. S., Janevic, T. M., Holzman, I. R., Yehuda, R., Landrigan, P. J. (2003). The World Trade Center disaster and intrauterine growth restriction. J. Amer. Med. Assoc. 290:595–596.

Bhate, S. & Wilkinson, S. (2006). Aetiology of learning disability. Psychiatry. 5:298–301.

Bird, A. (2007). Perceptions of epigenetics. Nature. 2007 447:396–398.

Blake-Gumbs, L. J. (2006). Global impact of lead poisoning in children and adults. In: Neuhauser, D. (ed.) Case Western Reserve University: Public Health Management & Policy. http://www.case.edu/med/epidbio/mphp439/Global_Impact_of_Lead_Poisoning.htm

Bobak, M. (2000). Outdoor air pollution, low birth weight, and prematurity. Environ. Health Perspect. 108:173–176.

Bobo, J. K., Klepinger, D. H., Dong, F. B. (2006). Changes in the prevalence of alcohol use during pregnancy among recent and at risk drinkers in the NLSY cohort. J. Women's Health. 15:1061–1070.

Bonilha, L., Cendes, F., Rorden, C., Eckert, M., Dalgalarrondo, P., Li, L. M., Steiner, C. E. (2008). Gray and white matter imbalance— Typical structural abnormality underlying classic autism? Brain Dev. 30:396–401.

Boog, G. (2004). Obstetrical complications and subsequent schizophrenia in adolescent and young adult offsprings: is there a relationship? Eur. J. Obstet. Gynecol. Reprod. Biol. 114:130–136.

Bookstein, F. L., Streissguth, A. P., Sampson, P. D., Connor, P. D., Barr, H. M. (2002). Corpus callosum shape and neuropsychological deficits in adult males with heavy fetal alcohol exposure. Neuroimage. 15:233–251.

Bor, W., Najman, J. M., Andersen, M. J., O'Callaghan, M., Williams, G. M., Behrens, B. C. (1997). The relationship between low family income and psychological disturbance in young children: an Australian longitudinal study. Aust. N. Z. J. Psychiatry. 31:664–675.

Borrell, L. N., Factor-Litvak, P., Wolff, M. S., Susser, E., Matte, T. D. (2004). Effect of socioeconomic status on exposures to polychlorinated biphenyls (PCBs) and dichlorodiphenyldichloroethylene (DDE) among pregnant African-American women. Arch. Environ. Health. 59:250–255.

Bouchard, T. J. Jr. & McGue, M. (2003). Genetic and environmental influences on human psychological differences. J. Neurobiol. 54:4–45.

Bouchard, T. J. Jr., Lykken, D. T., McGue, M., Segal, N. L., Tellegen, A. (1990). Sources of human psychological differences: the Minnesota study of twins reared apart. Science. 250:223–228.

Bove, F., Shim, Y., Zeitz, P. (2002). Drinking water contaminants and adverse pregnancy outcomes: a review. Environ. Health Perspect. 110(Suppl. 1):61–74.

Boyle, C. A., Decouflé, P., Yeargin-Allsopp, M. (1994). Prevalence and health impact of developmental disabilities in US children. Pediatrics. 93:399–403.

Boyle, C. A., Yeargin-Allsopp, M., Doernberg, N. S., Holmgreen, P., Murphy, C. C., Schendel, D. E. (1996). Prevalence of selected developmental disabilities in children 3–10 years of age: the Metropolitan Atlanta Developmental Disabilities Surveillance Program, 1991. Morb. Mort. Wkly. Rep. CDC Surveill Summ. 1996 45:1–14.

Bradley, S. J., Oliver, G. D., Chernick, A. B., Zucker, K. J. (1998). Experiment of nurture: ablatio penis at 2 months, sex reassignment at 7 months, and a psychosexual follow-up in young adulthood. Pediatrics. 102:e9.

Bransfield, R. C., Wulfman, J. S., Harvey, W. T., Usman, A. I. 2008. The association between tick-borne infections, Lyme borreliosis and autism spectrum disorders. Med. Hypotheses. 70:967–74.

Braun, J. M. & Lanphear, B. (2007). Comments on "Lead neurotoxicity in children: is prenatal exposure more important than postnatal exposure?". Acta Paediatr. 96:473.

Braun, J. M., Kahn, R. S., Froehlich, T., Auinger, P., Lanphear, B. P. (2006). Exposures to environmental toxicants and attention deficit hyperactivity disorder in U.S. children. Environ. Health Perspect. 114:1904–1909.

Brennan, P. A., Grekin, E. R., Mortensen, E. L., Mednick, S. A. (2002). Relationship of maternal smoking during pregnancy with criminal arrest and hospitalization for substance abuse in male and female adult offspring. Am. J. Psychiatry. 159:48–54.

Briellmann, R. S., Jackson, G. D., Torn-Broers, Y., Berkovic, S. F. (2001). Causes of epilepsies: insights from discordant monozygous twins. Ann. Neurol. 49:45–52.

Brooks, G. F., Carroll, K. C., Butel, J. S., Morse, S. A. (2007). *Medical Microbiology*. 24th edition. New York: McGraw-Hill.

Brown, A. S., Susser, E. S., Lin, S. P., Neugebauer, R., Gorman, J. M. (1995). Increased risk of affective disorders in males after second trimester prenatal exposure to the Dutch hunger winter of 1944–45. Br. J. Psychiatry. 166:601–606.

Brunton, P. J. & Russell, J. A. (2008). The expectant brain: adapting for motherhood. Nat. Rev. Neurosci. 9:11–25.

Bryan, W. J. (1924). Address to Seventh Day Adventists. Quoted from Milsted, D. (1999). *The Cassell Dictionary of Regrettable Quotations.* New York: Cassell.

Buccino, G. & Amore, M. (2008). Mirror neurons and the understanding of behavioural symptoms in psychiatric disorders. Curr. Opin. Psychiatry. 21:281–285.

Bulik, C. M., Sullivan, P. F., Tozzi, F., Furberg, H., Lichtenstein, P., Pedersen, N. L. Prevalence, heritability, and prospective risk factors for anorexia nervosa. Arch. Gen. Psychiatry. 2006 Mar 63(3):305–312.

Butterfield, F. (2002). Father steals best: crime in an American family. New York Times, August 21.

Caldwell, K. L., Jones, R., Hollowell, J. G. (2005). Urinary iodine concentration: United States National Health and Nutrition Examination Survey 2001–2002. Thyroid. 15:692–699.

Cameron, H. & Dayer, A. G. (2008). New interneurons in the adult neo-cortex: small, sparse, but significant? Biol. Psychiatry. 63:650–655.

Canby, V. (1988). Brotherly love, of sorts. New York Times, December 18, 1988.

Canfield, R. L., Henderson, C. R. Jr., Cory-Slechta, D. A., Cox, C., Jusko, T. A., Lanphear, B. P. (2003). Intellectual impairment in children with blood lead concentrations below 10 microg per deciliter. N. Engl. J. Med. 348:1517–1526.

Canino, G. & Alegria, M. (2008). Psychiatric diagnosis—is it universal or relative to culture? J. Child Psychol. Psychiatry. 49:237–250.

Cannon, T. D., Van Erp, T. G., Rosso, I. M., Huttunen, M., Lönnqvist, J., Pirkola, T., Salonen, O., Valanne, L., Poutanen, V. P., Standertskjöld-Nordenstam, C. G. (2002). Fetal hypoxia and structural brain abnormalities in schizophrenic patients, their siblings, and controls. Arch. Gen. Psychiatry. 59:35–41.

Caspi, A. & Silva, P. A. (1995). Temperamental qualities at age three predict personality traits in young adulthood: longitudinal evidence from a birth cohort. Child Dev. 66:486–98.

Castrogiovanni, P., Iapichino, S., Pacchierotti, C., Pieraccini, F. (1998). Season of birth in psychiatry. A review. Neuropsychobiol. 37:175–81.

CBC News. (2004). David Reimer: the boy who lived as a girl. May 10, 2004. http://www.cbc.ca/news/background/reimer/

Ceccanti, M., Alessandra-Spagnolo, P., Tarani, L., Luisa-Attilia, M., Chessa, L., Mancinelli, R., Stegagno, M., Francesco Sasso, G., Romeo, M., Jones, K. L., Robinson, L. K., Del Campo, M., Phillip-Gossage, J., May, P. A., Eugene-Hoyme, H. (2007). Clinical delineation of fetal alcohol spectrum disorders (FASD) in Italian children: comparison and contrast with other racial/ethnic groups and implications for diagnosis and prevention. Neurosci. Biobehav. Rev. 31:270–277.

Centers for Disease Control and Prevention. (2002). Fetal alcohol syndrome—Alaska, Arizona, Colorado, and New York, 1995–1997. Morb. Mort. Wkly. Rep. 51:433–435.

———. (2002b). Racial and ethnic disparities in infant mortality rates—60 largest U. S. cities, 1995–1998. Morb. Mort. Wkly. Rep. 51:329–332.

———. (2002c). Infant mortality and low birth weight among black and white infants—United States, 1980–2000. Morb. Mort. Wkly. Rep. 51:589–592.

———. (2004a). Cerebral palsy. http://www.cdc.gov/ncbddd/dd/cp3.htm#common

———. (2004b). Vision impairment. http://www.cdc.gov/ncbddd/dd/vision3.htm

———. (2004c). Hearing loss. http://www.cdc.gov/ncbddd/dd/hi3.htm

Centers for Disease Control and Prevention. (2004d) Alcohol consumption among women who are pregnant or who might become pregnant—United States, 2002. Morb. Mort. Wkly. Rep. 53:1178–1181

———. (2005). Blood lead levels—United States, 1999–2002. Morb. Mort. Wkly Rep. 54:513–516.

———. (2007). Interpreting and managing blood lead levels < 10 micrograms/dL in children and reducing childhood exposures to lead: recommendations of CDC's Advisory Committee on Childhood Lead Poisoning Prevention. Morb. Mort. Wkly. Rep. 56:#RR-8.

———. (2007b). Prevalence of autism spectrum disorder—Autism and Developmental Disabilities Monitoring Network, 14 sites, United States, 2002. Morb. Mort. Wkly. Rep. 56:12–28.

———. (2008). Epilepsy. http://www.cdc.gov/Epilepsy/

———. (2008b). Quickstats: mean gestational age, by plurality—United States, 2005. Morb. Mort. Wkly. Rep. 57:238.

———. (2008c). Update on overall prevalence of major birth defects—Atlanta, Georgia, 1978–2005. Morb. Mort. Wkly. Rep. 57:1–5.

Centers for Disease Control and Prevention/NCHS. (2008). Low birthweight: US/State, 1996–2004 (NVS). http://209.217.72.34/HDAA/TableViewer/tableView.aspx?ReportId=224

Centers for Disease Control and Prevention/NVSS. (2007). Fetal and perinatal mortality, United States, 2004. Natl. Vital Stat. Rep. 56(3):1–20.

Chambers, C. D., Hughes, S., Meltzer, S. B., Wahlgren, D., Kassem, N., Larson, S., Riley, E. P., Hovell, M. F. (2005). Alcohol consumption among low-income pregnant Latinas. Alcohol Clin. Exp. Res. 29:2022–2028.

Chen, A., Cai, B., Dietrich, K. N., Radcliffe, J., Rogan, W. J. (2007). Lead exposure, IQ, and behavior in urban 5- to 7-year-olds: does lead affect behavior only by lowering IQ? Pediatrics. 119:e650-e658.

Choi, B. H. (1988). The effects of methylmercury on the developing brain. Progr. Neurobiol. 32:447–470.

Claudio, L. (2002). The Hudson: a river runs through an environmental controversy. Environ. Health Perspect. 110:A184-A187.

Coggins, T. E., Timler, G. R., Olswang, L. B. (2007). A state of double jeopardy: impact of prenatal alcohol exposure and adverse environments on the social communicative abilities of school-age children with fetal alcohol spectrum disorder. Lang. Speech. Hear. Serv. Sch. 38:117–127.

Cohen-Bendahan, C. C., Van de Beek, C., Berenbaum, S. A. (2005). Prenatal sex hormone effects on child and adult sex-typed behavior: methods and findings. Neurosci. Biobehav. Rev. 29:353–384.

Colman, A. M. (2006). *A Dictionary of Psychology*. 2nd edition. New York: Oxford University Press.

Colugnati, F. A., Staras, S. A., Dollard, S. C., Cannon, M. J. (2007). Incidence of cytomegalovirus infection among the general population and pregnant women in the United States. BMC Infect. Dis. 2007 7:71.

Conley, D., Strully, K. W., Bennett, N. G. (2006). Twin differences in birth weight: the effects of genotype and prenatal environment on neonatal and post-neonatal mortality. Econ. Hum. Biol. 4:151–183.

Copeland, W. E., Miller-Johnson, S., Keeler, G., Angold, A., Costello, E. J. (2007). Childhood psychiatric disorders and young adult crime: a prospective, population-based study. Am. J. Psychiatry. 164:1668–1675.

Cornelius, M. D., Goldschmidt, L., DeGenna, N., Day, N. L. (2007). Smoking during teenage pregnancies: effects on behavioral problems in offspring. Nicotine Tob. Res. 9:739–750.

Correia, L. L. & Linhares, M. B. (2007). Maternal anxiety in the pre- and postnatal period: a literature review. Rev. Lat. Am. Enfermagem. 15:677–683.

Cortese, B. M., Moore, G. J., Bailey, B. A., Jacobson, S. W., Delaney-Black, V., Hannigan, J. H. (2006). Magnetic resonance and spectroscopic imaging in prenatal alcohol-exposed children: preliminary findings in the caudate nucleus. Neurotoxicol. Teratol. 28:597–606.

Costello, E. J., Compton, S. N., Keeler, G., Angold, A. (2003). Relationships between poverty and psychopathology: a natural experiment. J Am. Med. Assoc. 290:2023–2029.

Coussons-Read, M. E., Okun, M. L., Nettles, C. D. (2007). Psychosocial stress increases inflammatory markers and alters cytokine production across pregnancy. Brain Behav. Immun. 21:343–350.

Couzin, J. (2002). Quirks of fetal environment felt decades later. Science. 296:2167–2169.

Cowperthwaite, B., Hains, S. M., Kisilevsky, B. S. (2007). Fetal behavior in smoking compared to non-smoking pregnant women. Infant Behav. Dev. 30:422–430.

Craddock, N. & Owen, M. J. (2007). Rethinking psychosis: the disadvantages of a dichotomous classification now outweigh the advantages. World Psychiatry. 6:84–91.

Craddock, N., O'Donovan, M. C., Owen, M. J. (2006). Genes for schizophrenia and bipolar disorder? Implications for psychiatric nosology. Schizophr. Bull. 32:9–16.

Croen, L. A., Najjar, D. V., Fireman, B., Grether, J. K. (2007). Maternal and paternal age and risk of autism spectrum disorders. Arch. Pediatr. Adolesc. Med. 161:334–340.

Cuzon, V. C., Yeh, P. W., Yanagawa, Y., Obata, K., Yeh, H. H. (2008). Ethanol consumption during early pregnancy alters the disposition of tangentially migrating GABAergic interneurons in the fetal cortex. J. Neurosci. 28:1854–1864.

Daley, T. C., Whaley, S. E., Sigman, M. D., Espinosa, M. P., Neumann, C. (2003). IQ on the rise: the Flynn effect in rural Kenyan children. Psychol. Sci. 14:215–219.

Dalteg, A., Gustafsson, P., Levander, S. (1998). [Hyperactivity syndrome is common among prisoners. ADHD not only a pediatric psychiatric diagnosis] Lakartidningen 95(26–27):3078–3080. [Article in Swedish]

Daniels, J. L., Forssen, U., Hultman, C. M., Cnattingius, S., Savitz, D. A., Feychting, M., Sparen, P. (2008). Parental psychiatric disorders associated with autism spectrum disorders in the offspring. Pediatrics. 121:e1357-e1362.

Darwin, C. (1845). The voyage of the Beagle. In: Wilson, E. O. (ed.) *From So Simple a Beginning: The Four Great Books of Charles Darwin.* (2006). New York: W. W. Norton.

———. (1859). *On the Origin of Species.* Baltimore: Penguin Books, 1968.

———. (1879). *The Descent of Man, and Selection in Relation to Sex.* London: Penguin Books, 2004.

Davis, E. P., Glynn, L. M., Schetter, C. D., Hobel, C., Chicz-Demet, A., Sandman, C. A. (2007). Prenatal exposure to maternal depression and cortisol influences infant temperament. J. Am. Acad. Child Adolesc. Psychiatry. 46:737–746.

Dawkins, R. (1976). *The Selfish Gene.* New York: Oxford University Press.

Denno, D. W. (1990). *Biology and Violence.* New York: Cambridge University Press.

Desoto, M. C. & Hitlan, R. T. (2007). Blood levels of mercury are related to diagnosis of autism: a reanalysis of an important data set. J. Child Neurol. 22:1308–1311.

Detjen, M. G., Nieto, F. J., Trentham-Dietz, A., Fleming, M., Chasan-Taber, L. (2007). Acculturation and cigarette smoking among pregnant Hispanic women residing in the United States. Am. J. Public Health. 97:2040–2047.

Devlin, A. M., Cross, J. H., Harkness, W., Chong, W. K., Harding, B., Varga-Khadem, F., Neville, G. R. (2003). Clinical outcomes of hemispherectomy for epilepsy in childhood and adolescence. Brain. 126:556–566.

Devlin, B., Daniels, M., Roeder, K. (1997). The heritability of IQ. Nature. 388:468–471.

De Weerth, C., Van Hees, Y., Buitelaar, J. K. Prenatal maternal cortisol levels and infant behavior during the first 5 months. Early Hum. Dev. 74:139–151.

Dickens, W. T. & Flynn, J. R. (2001). Heritability estimates versus large environmental effects: the IQ paradox resolved. Psychol. Rev. 108:346–369.

Dipietro, J. A., Ghera, M. M., Costigan, K. A. (2008). Prenatal origins of temperamental reactivity in early infancy. Early Hum. Dev.84:569–575.

Dohrenwend, B. P., Levav, I., Shrout, P. E., Schwartz, S., Naveh, G., Link, B. G., Skodol, A. E., Stueve, A. (1992). Socioeconomic Status and Psychiatric Disorders: The Causation-Selection Issue. Science. 255:946–952.

Dörner, G., Götz, F., Rohde, W., Plagemann, A., Lindner, R., Peters, H., Ghanaati, Z. (2001). Genetic and epigenetic effects on sexual brain organization mediated by sex hormones. Neuro. Endocrinol. Lett. 22:403–409.

Downes, S. M. (2004). Heredity and heritability. In: Edward N. Zalta (ed.) *The Stanford Encyclopedia of Philosophy*. (Fall 2004 Edition). URL http://plato.stanford.edu/archives/fall2004/entries/heredity/

Drake, A. J., Tang, J. I., Nyirenda, M. J. (2007). Mechanisms underlying the role of glucocorticoids in the early life programming of adult disease. Clin. Sci. (Lond) 113:219–232.

Dudley, R. (2004). Ethanol, fruit ripening, and the historical origins of human alcoholism in primate frugivory. Integr. Comp. Biol. 44:315–323.

Dugdale, R. L. (1877). *The Jukes: A Study in Crime, Pauperism, Disease and Heredity.* New York: G.P. Putnam's Sons, 1891.

Ebrahim, S. H., Decouflé, P., Plakathodi, A. S. (2000). Combined tobacco and alcohol use by pregnant and reproductive-aged women in the United States. Obstetr. Gynecol. 96:767–771.

Edelman, G. M. (1987). *Neural Darwinism: The Theory of Neuronal Group Selection.* New York: Basic Books.

Edwards M. J. (2007). Hyperthermia in utero due to maternal influenza is an environmental risk factor for schizophrenia. Congenit. Anom. (Kyoto). 47:84–89.

Eichler, E. E. & Zimmerman, A. W. (2008). A hot spot of genetic instability in autism. New Engl. J. Med. 358:737–739.

Engel, S. M., Berkowitz, G. S., Wolff, M. S., Yehuda, R. (2005). Psychological trauma associated with the World Trade Center attacks and its effect on pregnancy outcome. Pediatr. Perinat. Epidemiol. 19:334–341.

Environmental Protection Agency. (2007). Environmental pollution and disease: draft report on the environment. http://www.epa.gov/Indicators/roe/html/roeHealthEn.htm

Farber, S. L. (1981). *Identical Twins Reared Apart: A Reanalysis.* New York: Basic Books.

Faroon, O., Jones, D., de Rosa, C. (2001). Effects of polychlorinated biphenyls on the nervous system. Toxicol. Ind. Health 16:305–333.

Fast, D. K. & Conry, J. (2004). The challenge of fetal alcohol syndrome in the criminal legal system. Addict. Biol. 9:161–166.

Favaro, A., Tenconi, E., Santonastaso, P. (2006). Perinatal factors and the risk of developing anorexia nervosa and bulimia nervosa. Arch. Gen. Psychiatry. 63:82–88.

Felton, J. S., Knize, M. G., Wu, R. W., Colvin, M. E., Hatch, F. T., Malfatti, M. A. (2007). Mutagenic potency of food-derived heterocyclic amines. Mutat. Res. 616:90–94.

Flick, L. H., Cook, C. A., Homan, S. M., McSweeney, M., Campbell, C., Parnell, L. (2006). Persistent tobacco use during pregnancy and the likelihood of psychiatric disorders. Am. J. Public Health. 96:1799–1807.

Floyd, R. L., Decouflé, P., Hungerford, D. W. (1999). Alcohol use prior to pregnancy recognition. Am. J. Prev. Med. 17:101–107.

Flynn, J. R. (1984). The mean IQ of Americans: massive gains. Psychol. Bull. 95:25–51.

———. (1987). Massive IQ gains in 14 nations. What IQ tests really mean. Psychol. Bull. 101:171–191.

———. (1998). IQ gains over time: toward finding the causes. In: Neisser, U. (ed.) *The Rising Curve: Long-Term Gains in IQ and Related Measures.* Washington, DC: American Psychological Association, pp. 25–66.

———. (2007). *What Is Intelligence?* New York: Cambridge University Press.

Fodor, J. (1998). The trouble with psychological Darwinism. London Review of Books, 22 January.

Fornoff, J. E., Easton, K., Wilson, T., Shen, T. (2007). Trends in the prevalence of birth defects in Illinois and Chicago, 1989–2004. Epidemiological Report Series 07–04, Springfield, IL., Illinois Department of Public Health, April 2007.

Fornoff, J. E., Wilson, T., Shen, T. (2007). Birth defects and other adverse pregnancy outcomes in Illinois 2001–2005. Epidemiological Report Series 07–06, Springfield, IL., Illinois Department of Public Health, June 2007.

Fox, D. J. & Druschel, C. M. (2003). Estimating prevalence of fetal alcohol syndrome (FAS): effectiveness of a passive birth defects registry system. Birth Defects Res. A Clin. Mol. Teratol. 67:604–608.

Freedman, A. M., Kaplan, H. I., Sadock, B. J. (1976). *Modern Synopsis of Comprehensive Textbook of Psychiatry/II.* 2nd edition. Baltimore: Williams & Wilkins.

Fumagalli, F., Molteni, R., Racagni, G., Riva, M. A. (2007). Stress during development: Impact on neuroplasticity and relevance to psychopathology. Prog. Neurobiol. 81:197–217.

Gale, C. R. & Martyn, C. N. (2004). Birth weight and later risk of depression in a national birth cohort. Br. J. Psychiatry. 184:28–33.

Galen. (1968) *De Usu Partium* [On the Usefulness of the Parts], Book 14, transl. May, M. T., Volume 2. Ithaca: Cornell University Press.

Gallagher, B. J. 3rd, McFalls, J. A. Jr., Jones, B. J., Pisa, A. M. (1999). Prenatal illness and subtypes of schizophrenia: the winter pregnancy phenomenon. J. Clin. Psychol. 55:915–922.

Geschwind, D. H. & Levitt, P. (2007). Autism spectrum disorders: developmental disconnection syndromes. Curr. Opin. Neurobiol. 17:103–11.

Gilbert, S. F. (2006). *Developmental Biology.* 8th edition. Sunderland, MA: Sinauer Associates.

Gillberg C. (1999). Neurodevelopmental processes and psychological functioning in autism. Dev. Psychopathol. 11:567–587.

Gillberg, C., Steffenburg, S., Schaumann, H. (1991). Is autism more common now than ten years ago? Br. J. Psychiatry. 158:403–409.

Glazier, A. M., Nadeau, J. H., Altman, T. J. (2002). Finding genes that underlie complex traits. Science. 298:2345–2349.

Gluckman, P. D., Hanson, M. A., Spencer, H. G., Bateson, P. (2005). Environmental influences during development and their later consequences for health and disease: implications for the interpretation of empirical studies. Proc. R. Soc. B 272:671–677.

Goodman, C. S. & Coughlin, B. C. (2000). The evolution of evo-devo biology. Proc. Natl. Acad. Sci. 97:4424–4425.

Goodwin, R. D., Keyes, K., Simuro, N. (2007). Mental disorders and nicotine dependence among pregnant women in the United States. Obstet. Gynecol. 109:875–883.

Gooren, L. (2006). The biology of human psychosexual differentiation. Horm. Behav. 50:589–601.

Gortmaker, S. L. (1979). The effects of prenatal care upon the health of the newborn. Am. J. Public Health. 69:653–660.

Gottesman, I. I. (1991). *Schizophrenia Genesis: The Origins of Madness*. New York: Freeman.

Gottesman, I. I. & Gould, T. D. (2003). The endophenotype concept in psychiatry: etymology and strategic intentions. Am. J. Psychiatry. 160:636–645.

Gould, E., Reeves, A. J., Graziano, M. S., Gross, C. G. (1999). Neurogenesis in the neocortex of adult primates. Science. 286:548–552.

Gould, J. B., Madan, A., Qin, C., Chavez, G. (2003). Perinatal outcomes in two dissimilar immigrant populations in the United States: a dual epidemiologic paradox. Pediatrics. 111(6 Pt 1):e676-e682.

Gould, S. J. (1996). *The Mismeasure of Man*. New York: W. W. Norton.

Gould, S. J. & Lewontin, R. C. (1979). The spandrels of San Marco and the Panglossian paradigm: a critique of the adaptationist programme. Proc. Roy. Soc. London, Series B 205:581–598.

Grandjean, P. (2008). Late insights into early origins of disease. Basic & Clin. Pharmacol. Toxicol. 102:94–99.

Grandjean, P., Bellinger, D., Bergman, A., Cordier, S., Davey-Smith, G., Eskenazi, B., Gee, D., Gray, K., Hanson, M., van den Hazel, P., Heindel, J. J., Heinzow, B., Hertz-Picciotto, I., Hu, H., Huang, T. T., Jensen, T. K., Landrigan, P. J., McMillen, I. C., Murata, K., Ritz, B., Schoeters, G., Skakkebaek, N. E., Skerfving, S., Weihe, P. (2008). The Faroes statement: human health effects of developmental exposure to chemicals in our environment. Basic Clin. Pharmacol. Toxicol. 102:73–75.

Grandjean, P., Landrigan, P. J. (2006). Developmental neurotoxicity of industrial chemicals. Lancet 368:2167–2178.

Gray, K. A., Klebanoff, M. A., Brock, J. W., Zhou, H., Darden, R., Needham, L., Longnecker, M. P. (2005). In utero exposure to background levels of polychlorinated biphenyls and cognitive functioning among school-age children. Am. J. Epidemiol. 162:17–26.

Griffiths, C., McGartland, A., Miller, M. (2007). A comparison of the monetized impact of IQ decrements from mercury emissions. Environ. Health Perspect. 2007 115:841–8477.

Grigg, J. (2004). Environmental toxins: their impact on children's health. Archiv. Disease in Childhood. 89:244–250.

Grove, E. A. (2008). Organizing the source of memory. Science. 319:288–289.

Habek, D. (2007). Effects of smoking and fetal hypokinesia in early pregnancy. Arch. Med. Res. 38:864–8677.

Habek, D., Habek, J. C., Ivanisevic, M., Djelmis, J. (2002). Fetal tobacco syndrome and perinatal outcome. Fetal Diagn. Ther. 17:367–371.

Habib, M. (2000). The neurological basis of developmental dyslexia: an overview and working hypothesis. Brain. 123:2373–2399.

Haier, R. J., Jung, R. E., Yeo, R. A., Head, K., Alkire, M. T. (2004). Structural brain variation and general intelligence. Neuroimage. 23:425–433.

Hall, L. S. & Love, C. T. (2003). Finger-length ratios in female monozygotic twins discordant for sexual orientation. Arch. Sexual Behav. 32:23–28.

Hamlin, J. K., Wynn, K., Bloom, P. (2007). Social evaluation by preverbal infants. Nature. 450:557–560.

Hampton, T. (2004). Fetal environment may have profound long-term consequences for health. J. Amer. Med. Assoc. 292:1285–1286.

Hanson, M. A., Gluckman, P. D. (2008). Developmental origins of health and disease: new insights. Basic Clin. Pharmacol. Toxicol. 102:90–93.

Hardell, L., Lindström, G., Van Bavel, B. (2002). Is DDT exposure during fetal period and breast-feeding associated with neurological impairment? Environ. Res. 88:141–144.

Hardman, J. G., Limbird, L. E., Molinoff, P. B., Ruddon, R. W., Gilman, A. G. (1996) *Goodman & Gilman's The Pharmacological Basis of Therapeutics.* 9th edition. New York: McGraw-Hill.

Harmon, A. (2007). The DNA age: in DNA era, new worries about prejudice. New York Times, November 11, 2007.

Harper, L. V. (2005). Epigenetic inheritance and the intergenerational transfer of experience. Psychol Bull. 131:340–360.

Harrison, P. J. & Weinberger, D. R. (2005). Schizophrenia genes, gene expression, and neuropathology: on the matter of their convergence. Mol. Psychiatry. 10:40–68.

Harvey, C. D. & Svoboda, K. (2007). Locally dynamic synaptic learning rules in pyramidal neuron dendrites. Nature. 450:1195–1200.

Haustein, K. O. (2006). Smoking and poverty. Eur. J. Cardiovasc. Prev. Rehabil. 13:312–318.

Hellström-Lindahl, E. & Nordberg, A. (2002). Smoking during pregnancy: a way to transfer the addiction to the next generation? Respiration. 69:289–293.

Herrnstein, R. J. & Murray, C. (1994). *The Bell Curve: Intelligence and Class Structure in American Life.* New York: Free Press.

Hershberger, S. L. & Segal, N. L. (2004). The cognitive, behavioral, and personality profiles of a male monozygotic triplet set discordant for sexual orientation. Arch. Sex. Behav. 33:497–514.

Hertz-Picciotto, I., Park, H. Y., Dostal, M., Kocan, A., Trnovec, T., Sram, R. (2008). Prenatal exposures to persistent and non-persistent organic compounds and effects on immune system development. Basic Clin. Pharmacol. Toxicol. 102:146–54.

Hines, M. (2006). Prenatal testosterone and gender-related behavior. Eur. J. Endocrinol. 155:S115-S121.

Honeycutt, R. L. (2008). Small changes, big results: evolution of morphological discontinuity in mammals. J. Biol. 7:9. http://jbiol.com/content/7/3/9

Hong, E. J., West, A. E., Greenberg, M. E. (2005). Transcriptional control of cognitive development, Current Opinion in Neurobiology 15:21–28.

Howes, O. D., McDonald, C., Cannon, M., Arseneault, L., Boydell, J., Murray, R. M. (2004). Pathways to schizophrenia: the impact of environmental factors. Int. J. Neuropsychopharmacol. (7 Suppl 1):S7-S13.

Hu, H., Téllez-Rojo, M. M., Bellinger, D., Smith, D., Ettinger, A. S., Lamadrid-Figueroa, H., Schwartz, J., Schnaas, L., Mercado-García, A., Hernández-Avila, M. (2006). Fetal lead exposure at each stage of pregnancy as a predictor of infant mental development. Environ. Health Perspect. 114:1730–1735.

Huang, L., Sauve, R., Birkett, N., Fergusson, D., Van Walraven, C. (2008). Maternal age and risk of stillbirth: a systematic review. Can. Med. Assoc. J. 178:165–172.

Huizink, A. C., Bartels, M., Rose, R. J., Pulkkinen, L., Eriksson, C. J., Kaprio, J. (2008). Chernobyl exposure as stressor during pregnancy and hormone levels in adolescent offspring. J. Epidemiol. Community Health. 62(4):e5.

Huizink, A. C., Dick, D. M., Sihvola, E., Pulkkinen, L., Rose, R. J., Kaprio, J. (2007). Chernobyl exposure as stressor during pregnancy and behaviour in adolescent offspring. Acta Psychiatr. Scand. 116:438–446.

Huizink, A. C., Mulder, E. J., Buitelaar, J. K. (2004). Prenatal stress and risk for psychopathology: specific effects or induction of general susceptibility? Psychol. Bull. 130:115–142.

Hwang, L. (2007). Environmental stressors and violence: lead and polychlorinated biphenyls. Rev. Environ. Health 22:313–328.

Iacoboni, M. & Dapretto, M. (2006). The mirror neuron system and the consequences of its dysfunction. Nat. Rev. Neurosci. 7:942–951.

Innis, S. M. (2007). Fatty acids and early human development. Early Hum. Dev. 83:761–766.

Istvan, J. (1986). Stress, anxiety, and birth outcomes: a critical review of the evidence. Psychol. Bull. 100:331–348.

Jablonka, E. & Lamb, M. (2005). *Evolution in Four Dimensions*. Cambridge, MA: The MIT Press.

———. (2007). Précis of Evolution in Four Dimensions. Behav. Brain. Sci. 30:353–392.

Jacobs, N., Van Gestel, S., Derom, C., Thiery, E., Vernon, P., Derom, R., Vlietinck, R. (2001). Heritability estimates of intelligence in twins: effect of chorion type. Behav. Genet. 31:209–217.

Jacobson, J. L. & Jacobson, S. W. (1996). Intellectual impairment in children exposed to polychlorinated biphenyls in utero. N. Engl. J. Med. 335:783–789.

Jacobson, S. W., Stanton, M. E., Molteno, C. D., Burden, M. J., Fuller, D. S., Hoyme, H. E., Robinson, L. K., Khaole, N., Jacobson, J. L. (2008). Impaired eyeblink conditioning in children with fetal alcohol syndrome. Alcohol Clin. Exp. Res. 32:365–372.

James W. H. (2005). Biological and psychosocial determinants of male and female human sexual orientation. J. Biosoc. Sci. 2005 37:555–567

Jansson, T. & Powell, T. L. (2007). Role of the placenta in fetal programming: underlying mechanisms and potential interventional approaches. Clin. Sci. (Lond). 113:1–13.

Jensen, A. R. (1969). How much can we boost IQ and scholastic achievement? Harvard Educ. Rev. 39:1–123.

Johnson, M. H. (2005). Sensitive periods in functional brain development: problems and prospects. Dev. Psychol. 46:287–292.

Jones, K. L. & Smith, D. W. (1973). Recognition of the fetal alcohol syndrome in early infancy. Lancet. November 3. 2:999–1001.

Jones, K. L., Smith, D. W., Ulleland, C. N., Streissguth, A. P. (1973). Pattern of malformation in offspring of chronic alcoholic mothers. Lancet. June 9. 1:1267–1271.

Joseph, J. (2001). Separated twins and the genetics of personality differences: a critique. Amer. J. Psychol. 114:1–30.

———. (2002). Twin studies in psychiatry and psychology: science or pseudoscience? Psychiatr. Q. 73:71–82.

Judson, H. F. (2001). Talking about the genome. Nature. 409:769.

Jauniaux, E. & Burton, G. J. (2007). Morphological and biological effects of maternal exposure to tobacco smoke on the feto-placental unit. Early Hum. Dev. 83:699–706.

Kajantie, E. (2006). Fetal origins of stress-related adult disease. Ann. N. Y. Acad. Sci. 1083:11–27.

Kamin, L. J. (1974). *The Science and Politics of IQ*. Potomac, MD: Lawrence Erlbaum Associates.

Kanner, L. (1943). Autistic disturbances of affective contact. Nerv. Child. 2:217–250.

Kaplan, G. & Rogers, L. J. (1994). Race and gender fallacies: the paucity of biological determinist explanations of difference. In: Tobach,

E. & Rosoff, B. (ed.) *Challenging Racism and Sexism. Alternatives to Genetic Explanations.* New York: Feminist Press. pp. 65–92.

Kaplan, G. & Rogers, L. J. (2003). *Gene Worship.* New York: Other Press.

Kempthorne, O. (1997). Heritability: uses and abuses. Genetica. 99:109–112.

Kenneally, C. (2006). The deepest cut. New Yorker. July 3, 2006.

Kenneson, A. & Cannon, M. J. (2007). Review and meta-analysis of the epidemiology of congenital cytomegalovirus (CMV) infection. Rev. Med. Virol. 17:253–276.

Kerjan, G. & Gleeson, J. G. (2007). Genetic mechanisms underlying abnormal neuronal migration in classical lissencephaly. Trends Genet. 23:623–630.

Kessler, R. C. & Wang, P. S. (2008). The descriptive epidemiology of commonly occurring mental disorders in the United States. Annu. Rev. Public Health. 29:115–129.

Kessler, R. C., Foster, C. L., Saunders, W. B., Stang, P. E. (1995). Social consequences of psychiatric disorders, I: Educational attainment. Am. J. Psychiatry. 152:1026–1032.

Khashan, A. S., Abel, K. M., McNamee, R., Pedersen, M. G., Webb, R. T., Baker, P. N., Kenny, L. C., Mortensen, P. B. (2008). Higher risk of offspring schizophrenia following antenatal maternal exposure to severe adverse life events. Arch. Gen. Psychiatry. 65:146–152.

Khattak, S., K-Moghtader, G., McMartin, K., Barrera, M., Kennedy, D., Koren, G. (1999). Pregnancy outcome following gestational exposure to organic solvents: a prospective controlled study. J. Amer. Med. Assoc. 281:1106–1109.

Kierans, W. J., Joseph, K. S., Luo, Z. C., Platt, R., Wilkins, R., Kramer, M. S. (2008). Does one size fit all? The case for ethnic-specific standards of fetal growth. BMC. Pregnancy Childbirth. 8:1–9.

Kinder, B. N. (1997). Eating disorders. In: Turner, S. M. & Hersen, M. (eds.) *Adult Psychopathology and Diagnosis*. New York: John Wiley & Sons. pp. 465–482.

King, S., Barr, R. G., Brunet, A., Saucier, J. F., Meaney, M., Woo, S., Chanson, C. (2000). [The ice storm: an opportunity to study the effects of prenatal stress on the baby and the mother.] Sante Ment. Que. 25:163–185. French.

Kodituwakku, P. W. (2007). Defining the behavioral phenotype in children with fetal alcohol spectrum disorders: a review. Neurosci. Biobehav. Rev. 31:192–201.

Koentges, G. (2008). Evolution of anatomy and gene control. Nature. 451:658–663.

Korkman, M., Granström, M-L., Kantola-Sorsa, E., Gaily, E., Paetau, R., Liukkonen, E., Boman, P-A., Blomstedt, G. (2005). Two-year follow-up of intelligence after pediatric epilepsy surgery. Pediatr. Neurol. 33:173–178.

Kramer, M. S., Ananth, C. V., Platt, R. W., Joseph, K. S. (2006). US Black vs White disparities in foetal growth: physiological or pathological? Int. J. Epidemiol. 35:1187–1195.

Landrigan, P. J., Sonawane, B., Butler, R. N., Trasande, L., Callan, R., Droller, D. (2005). Early environmental origins of neurodegenerative disease in later life. Environ. Health Perspect. 113:1230–1233.

Lanphear, B. P., Hornung, R., Khoury, J., Yolton, K., Baghurst, P., Bellinger, D. C., Canfield, R. L., Dietrich, K. N., Bornschein, R., Greene, T., Rothenberg, S. J., Needleman, H. L., Schnaas, L., Wasserman, G., Graziano, J., Roberts, R. (2005). Low-level environmental lead exposure and children's intellectual function: an international pooled analysis. Environ. Health Perspect. 113:894–899.

Laraque, D. & Trasande, L. (2005). Lead poisoning: successes and 21st century challenges. Pediatrics Rev. 26:435–443.

Laucht, M. & Schmidt, M. H. (2004). [Maternal smoking during pregnancy: risk factor for ADHD in the offspring?] Z. Kinder Jugendpsychiatr. Psychother. 32:177–185. German.

Layzer, D. (1974). Heritability analyses of IQ scores: science or numerology? Science 183:1259–1266.

Lerner, R. (2006). Another nine-inch nail for behavioral genetics! Hum. Dev. 49:336–342.

Levin, R., Brown, M. J., Kashtock, M. E., Jacobs, D. E., Whelan, E. A., Rodman, J., Schock, M. R., Padilla, A., Sinks, T. (2008). Lead exposures in U.S. Children, 2008: implications for prevention. Environ. Health Perspect. 116:1285–1293. Levine, A., Zagoory-Sharon, O., Feldman, R., Lewis, J. G., Weller, A. (2007). Measuring cortisol in human psychobiological studies. Physiol. Behav. 90:43–53.

Lewis, B. A., Kirchner, H. L., Short, E. J., Minnes, S., Weishampel, P., Satayathum, S., Singer, L. T. (2007). Prenatal cocaine and tobacco effects on children's language trajectories. Pediatrics. 120:e78–85.

Lewontin, R. C. (1974). The analysis of variance and the analysis of causes. Amer. J. Hum. Genet. 26:400–411.

———. (1982). *Human Diversity.* New York: Scientific American Books.

———. (2000). *The Triple Helix: Gene, Organism, and Environment.* Cambridge, MA: Harvard University Press.

Lewontin, R. C., Rose, S., Kamin, L. J. (1984). *Not In Our Genes: Biology, Ideology, and Human Nature.* New York: Pantheon.

Lidsky, T. I. & Schneider, J. S. (2003). Lead neurotoxicity in children; basic mechanisms and clinical correlates. Brain. 126:5–19.

————. (2005). Autism and autistic symptoms associated with childhood lead poisoning. J. Appl. Res. 5:80–87.

London, E. A. & Etzel, R. A. (2000). The environment as an etiologic factor in autism: a new direction for research. Environ. Health Perspect. (108 Suppl. 3):401–404.

Lumey L. H. & Van Poppel F. W. (1994). The Dutch famine of 1944–45: mortality and morbidity in past and present generations. Soc. Hist. Med. 7:229–46.

Machin, G. A. (1996). Some causes of genotype and phenotype discordances in monozygotic twin pairs. Am. J. Med. Genetics. 61:216–228.

Malamitsi-Puchner, A. & Boutsikou, T. (2006). Adolescent pregnancy and perinatal outcome. Pediatr. Endocrinol. Rev. (3 Suppl 1):170–171.

Mameli, M. & Bateson, P. (2006). Innateness and the sciences. Biology and Philosophy 21:155–188.

Mandler, G. (2001). Apart from genetics: what makes monozygotic twins similar? J. Mind & Behav. 22:147–159.

Mangale, V. S., Hirokawa, K. E., Satyaki, P. R., Gokulchandran, N., Chikbire, S., Subramanian, L., Shetty, A. S., Martynoga, B., Paul, J., Mai, M. V., Li, Y., Flanagan, L. A., Tole, S., Monuki, E. S. (2008). Lhx2 selector activity specifies cortical identity and suppresses hippocampal organizer fate. Science. 319:304–309.

March, W. (1954). *The Bad Seed*. New York: Dell.

Martin, R. D. (2007). The evolution of human reproduction: a primatological perspective. Am. J. Phys. Anthropol. (Suppl. 45):59–84.

Marx, C. E. & Lieberman, J. A. (1998). Psychoneuroendocrinology of schizophrenia. Psychiatr. Clin. North Am. 21:413–434.

Mastorakos, G. & Ilias, I. (2003). Maternal and fetal hypothalamic-pituitary-adrenal axes during pregnancy and postpartum. Ann. N. Y. Acad. Sci. 997:136–149.

Mattson, S. N., Riley, E. P., Jernigan, T. L., Ehlers, C. L., Delis, D. C., Jones, K. L., Stern, C., Johnson, K. A., Hesselink, J. R., Bellugi, U. (1992). Fetal alcohol syndrome: a case report of neuropsychological, MRI and EEG assessment of two children. Alcohol Clin. Exp. Res. 16:1001–1003.

May, P. A. & Gossage, J. P. (2001). Estimating the prevalence of fetal alcohol syndrome: a summary. Alcohol Res. Health. 25:159–167.

May P. A., Fiorentino, D., Phillip-Gossage, J., Kalberg, W. O., Eugene-Hoyme, H., Robinson, L. K., Coriale, G., Jones, K. L., Del Campo, M., Tarani, L., Romeo, M., Kodituwakku, P. W., Deiana, L., Buckley, D., Ceccanti, M. (2006). Epidemiology of FASD in a province in Italy: Prevalence and characteristics of children in a random sample of schools. Alcohol Clin. Exp. Res. 30:1562–1575.

McBarnette, L. (1987). Women and poverty: the effects on reproductive status. Women. Health. 12:55–81.

McConaghy, N. & Blaszczynski, A. (1980). A pair of monozygotic twins discordant for homosexuality: sex-dimorphic behavior and penile volume responses. Arch. Sex. Behav. 9:123–131.

McDaniel, M. A. (2006). Estimating state IQ: measurement challenges and preliminary correlates. Intelligence. 34:607–619.

McGowan, P. O., Sasaki, A., Huang, T. C. T., Unterberger, A., Suderman, M., Ernst, C., Meaney, M. J., Turecki, G., Szyf, M. (2008). Promoter-wide hypermethylation of the ribosomal RNA gene promoter in the suicide brain. PLoS ONE 3: e2085 doi:10.1371/journal.pone.0002085

McLachlan, J. A., Simpson, E., Martin, M. (2006). Endocrine disrupters and female reproductive health. Best Pract. Res. Clin. Endocrinol. Metab. 20:63–75.

Mcleod, J. D. & Shanahan, M. J. (1996). Trajectories of poverty and children's mental health. J. Health Soc. Behav. 37:207–220.

McPhillips, M. & Jordan-Black, J. A. (2007). The effect of social disadvantage on motor development in young children: a comparative study. J. Child Psychol. Psychiatry. 48:1214–1222.

Meaney, M. J. (2001). Maternal care, gene expression, and the transmission of individual differences in stress reactivity across generations. Annu. Rev. Neurosci. 24:1161–1192.

Mednick, S. A., Machon, R. A., Huttunen, M. O., Bonett, D. (1988). Adult schizophrenia following prenatal exposure to an influenza epidemic. Arch. Gen. Psychiatry. 45:189–192.

Meisenberg, G., Lawless, E., Lambert, E., Newton, A. (2005). The Flynn effect in the Caribbean: generational change of cognitive test performance in Dominica. Mankind. Q. 46:29–69.

Merlot, E., Couret, D., Otten, W. (2008). Prenatal stress, fetal imprinting and immunity. Brain Behav. Immun. 22:42–51.

Mervis, C. A., Yeargin-Allsopp, M., Winter, S., Boyle, C. (2000). Aetiology of childhood vision impairment, metropolitan Atlanta, 1991–93. Paediatr. Perinat. Epidemiol. 14:70–77.

Metacritic. (2008). http://www.metacritic.com/film/titles/rainman

Michel, A., Mormont, C., Legros, J. J. (2001). A psycho-endocrinological overview of transsexualism. Eur. J. Endocrinol. 145:365–376.

Michigan Department of Community Health.(2008). Birth defects prevalence and mortality tables. http://www.mdch.state.mi.us/pha/osr/BirthDefects/summary.asp

Mielke, H. W. (1999). Lead in the inner-cities, Am. Sci. 87:62–73.

Millichap, J. G. (2008). Etiologic Classification of Attention-Deficit/Hyperactivity Disorder. Pediatrics. 121:e358-e365

Ministry of the Environment, Government of Japan. (2002). Minamata disease: the history and measures. http://www.env.go.jp/en/chemi/hs/minamata2002/ch2.html

Minshew N. J. & Williams, D. L. (2007). The new neurobiology of autism: cortex, connectivity, and neuronal organization. Arch. Neurol. 64:945–950.

Molina, J. C., Spear, N. E., Spear, L. P., Mennella, J. A., Lewis, M. J. (2007). The International society for developmental psychobiology 39th annual meeting: Alcohol and development: beyond fetal alcohol syndrome. Dev. Psychobiol. 49:227–242.

Money, J. (1975). Ablatio penis: normal male infant sex-reassigned as a girl. Arch. Sex. Behav. 4:65–71.

Montgomery, S. M., Ehlin, A., Ekbom, A. (2005). Smoking during pregnancy and bulimia nervosa in offspring. J. Perinat. Med. 33:206–11.

Muhle, R., Trentacoste, S. V., Rapin, I. (2004). The genetics of autism. Pediatrics. 113:e472–486.

Muñoz-Tuduri, M. & Garcia-Moro, C. (2008) Season of birth affects short- and long-term survival. Am. J. Phys. Anthropol. 135:462–468.

Munsinger, H. (1977). The identical twin transfusion syndrome: a source of error in estimating IQ resemblance and heritability. Ann. Hum. Genet. 40:307–321.

Nadarajah, B., Alifragis, P., Wong, R. O., Parnavelas, J. G. (2003). Neuronal migration in the developing cerebral cortex: observations based on real-time imaging. Cerebral Cortex. 13:607–611.

Nafee, T. M., Farrell, W. E., Carroll, W. D., Fryer, A. A., Ismail, K. M. (2008) Epigenetic control of fetal gene expression. Brit. J. Obstetr. Gynecol. 115:158–68.

Nathanielsz, P. W. (1999). *Life in the Womb: The Origin of Health and Disease*. Ithaca, New York: Promethean Press.

————. (2000). Fetal programming: how the quality of fetal life alters biology for a lifetime. NeoReviews. 1:e126-e131.

Needleman, H. L., Gunnoe, C., Leviton, A., Reed, R., Peresie, H., Maher, C., Barrett, P. (1979). Deficits in psychologic and classroom performance of children with elevated dentine lead levels. N. Engl J Med. 300:689–695.

Needleman, H. L., Riess, J. A., Tobin, M. J., Biesecker, G. E., Greenhouse, J. B. (1996). Bone lead levels and delinquent behavior. J. Am. Med. Assoc. 275:363–369.

Neisser, U. (ed.) (1998). *The Rising Curve: Long-Term Gains in IQ and Related Measures*. Washington, DC: American Psychological Association.

Nelkin, D. & Lindee, M. S. (2004). *The DNA Mystique*. Ann Arbor, MI: University of Michigan Press.

Nevin, R. (2000). How lead exposure relates to temporal changes in IQ, violent crime, and unwed pregnancy. Environ. Res. 83:1–22.

Nevin, R. (2007). Understanding international crime trends: the legacy of preschool lead exposure. Environ. Res. 104:315–336.

Nierenberg, D. W., Nordgren, R. E., Chang, M. B., Siegler, R. W., Blayney, M. B., Hochberg, F., Toribara, T. Y., Cernichiari, E., Clarkson, T. (1998). Delayed cerebellar disease and death after accidental exposure to dimethylmercury. N. Engl. J. Med. 338: 1672–1676.

Nigg, J. T., Knottnerus, G. M., Martel, M. M., Nikolas, M., Cavanagh, K., Karmaus, W., Rappley, M. D. (2008). Low blood lead levels associated with clinically diagnosed attention-deficit/hyperactivity disorder and mediated by weak cognitive control. Biol. Psychiatry. 63:325–331.

O'Connor, M. J. & Whaley, S. E. (2003). Alcohol use in pregnant low-income women. J. Stud. Alcohol. 64:773–783.

O'Connor, T. G., Ben-Shlomo, Y., Heron, J., Golding, J., Adams, D., Glover, V. (2007). Prenatal anxiety predicts individual differences in cortisol in pre-adolescent children. Biol. Psychiatry. 58:211–217.

Olney, J. (2004). Fetal alcohol syndrome at the cellular level. Addict. Biol. 9:137–149.

Olson, R. K. (2006). Genes, environment, and dyslexia. Ann. Dyslexia. 56:205–238.

Opler, M. G. & Susser, E. S. (2005). Fetal environment and schizophrenia. Environ. Health Perspect. 113:1239–1242.

Opler, M. G., Brown, A. S., Graziano, J., Desai, M., Zheng, W., Schaefer, C., Factor-Litvak, P., Susser, E. S. (2004). Prenatal lead exposure, delta-aminolevulinic acid, and schizophrenia. Environ. Health Perspect. 112:548–552.

Ouellet-Morin, I., Boivin, M., Dionne, G., Lupien, S. J., Arsenault, L., Barr, R. G., Pérusse, D., Tremblay, R. E. (2008). Variations in heritability of cortisol reactivity to stress as a function of early familial adversity among 19-month-old twins. Arch. Gen. Psychiatry. 65:211–218.

Palomo, T., Archer, T., Beninger, R. J., Kostrzewa, R. M. (2004). Gene-environment interplay in neurogenesis and neurodegeneration. Neurotox. Res. 6:415–434.

Paneth, N., Korzeniewski, S., Hong, T. (2005). The Role of the Intrauterine and Perinatal Environment in Cerebral Palsy. NeoReviews. 6:e133-e140.

Patterson, P. H. (2007). Maternal effects on schizophrenia risk. Science. 318:576–577.

Patton, G. C., Coffey, C., Carlin, J. B., Olsson, C. A., Morley, R. (2004). Prematurity at birth and adolescent depressive disorder. Br. J. Psychiatry. 184:446–447

Paulesu, E., Démonet, J. F., Fazio, F., McCrory, E., Chanoine, V., Brunswick, N., Cappa, S. F., Cossu, G., Habib, M., Frith, C. D.,

Frith, U. (2001). Dyslexia: cultural diversity and biological unity. Science. 291:2165–2167.

———. (2001b). Defining dyslexia—response. Science. 292:1300–1301.

Pennisi, E. (2007). DNA study forces rethink of what it means to be a gene. Science. 316:1556–1557.

Perera, F., Hemminki, K., Jedrychowski, W., Whyatt, R., Campbell, U., Hsu, Y, Santella, R, Albertini, R, O'Neill, J. P. (2002). In utero DNA damage from environmental pollution is associated with somatic gene mutation in newborns. Cancer Epidemiol. Biomarkers Prev. 2002 11:1134–7.

Perera, F. P., Rauh, V., Whyatt, R. M., Tang, D., Tsai, W. Y., Bernert, J. T., Tu, Y. H., Andrews, H., Barr, D. B., Camann, D. E., Diaz, D., Dietrich, J., Reyes, A., Kinney, P. L. (2005). A summary of recent findings on birth outcomes and developmental effects of prenatal ETS, PAH, and pesticide exposures. Neurotoxicol. 26:573–587.

Pesonen, A. K., Räikkönen, K., Kajantie, E., Heinonen, K., Strandberg, T. E., Järvenpää, A. L. (2006). Fetal programming of temperamental negative affectivity among children born healthy at term. Dev. Psychobiol. 48:633–643.

Petronis, A. (2003). Epigenetics and bipolar disorder: new opportunities and challenges. Am. J. Med. Genet. C Semin. Med. Genet. 123:65–75.

———. (2004). The origin of schizophrenia: genetic thesis, epigenetic antithesis, and resolving synthesis. Biol. Psychiatry. 55:965–970.

Petronis, A., Gottesman, I. I., Kan, P., Kennedy, J. L., Basile, V. S., Paterson, A. D., Popendikyte, V. (2003). Monozygotic twins exhibit numerous epigenetic differences: clues to twin discordance? Schizophr. Bull. 29:169–178.

Phillips, D. I. & Jones, A. (2006). Fetal programming of autonomic and HPA function: do people who were small babies have enhanced stress responses? J. Physiol. 572(Pt 1):45–50.

Platt, S. A. & Bach, M. (1997). Uses and misinterprations of genetics in psychology. Genetica. 99:135–143.

Pollak, R. (1997). *The Creation of Dr. B.* New York: Simon & Schuster.

Prescott, C. A., Johnson, R. C., McArdle, J. J. (1999). Chorion type as a possible influence on the results and interpretation of twin study data. Twin Res. 2:244–249.

Previc, F. H. (2007). Prenatal influences on brain dopamine and their relevance to the rising incidence of autism. Med. Hypotheses. 68:46–60.

Procopio, M. & Marriott, P. K. (1998). Seasonality of birth in epilepsy: a Danish study. Acta Neurol. Scand. 98:297–301.

———. (2007). Intrauterine hormonal environment and risk of developing anorexia nervosa. Arch. Gen. Psychiatry. 64:1402–1407.

Procopio, M., Marriott, P. K., Williams, P. (1997). Season of birth: aetiological implications for epilepsy. Seizure. 6:99–105.

Pulsifer, M. B., Brandt, J., Salorio, C. F., Vining, E. P., Carson, B. S., Freeman, J. M. (2004). The cognitive outcome of hemispherectomy in 71 children. Epilepsia. 45:243–254.

Purves, D. (2007). *Neuroscience.* 4th edition. Sunderland, MA: Sinauer Associates.

Rabkin, J. G. & Struening, E. L. (1976). Life events, stress, and illness. Science. 194:1013–1020.

Rahman, Q. (2005). The neurodevelopment of human sexual orientation. Neurosci. Biobehav. Rev. 29:1057–1066.

Raison, C. L. & Miller, A. H. (2003). When not enough is too much: the role of insufficient glucocorticoid signaling in the pathophysiology of stress-related disorders. Am. J. Psychiatry. 160:1554–1565.

Rakic, P. & Lombroso, P. J. (1998). Development of the cerebral cortex: forming the cortical structure. J. Am. Acad. Child Adolesc. Psychiatry. 37:116–117.

Rapoport, J. L., Addington, A. M., Frangou, S., Psych, M. R. (2004). The neurodevelopmental model of schizophrenia: update 2005. Mol. Psychiatry. 10:434–449.

Rasmussen, C. (2005). Executive functioning and working memory in fetal alcohol spectrum disorder. Alcohol. Clin. Exp. Res. 29:1359–1367.

Rees S. & Harding R. (2004). Brain development during fetal life: influences of the intra-uterine environment. Neurosci. Lett. 2004 361:111–114.

Rees, S. & Inder, T. (2005). Fetal and neonatal origins of altered brain development. Early. Hum. Dev. 81:753–761.

Rees, S., Harding, R., Walker, D. (2008). An adverse intrauterine environment: implications for injury and altered development of the brain. Internatl. J. Dev. Neurosci. 26:3–11.

Reik, W., Dean, W., Walter, J. (2001). Epigenetic reprogramming in mammalian development. Science. 293:1089–1093.

Relier, J-P. (2001). Influence of maternal stress on fetal behavior and brain development. Biol. Neonate. 79:168–171.

Retz, W., Retz-Junginger, P., Hengesch, G., Schneider, M., Thome, J., Pajonk, F. G., Salahi-Disfan, A., Rees, O., Wender, P. H., Rösler, M. (2004). Psychometric and psychopathological characterization of young male prison inmates with and without attention deficit/hyperactivity disorder. Eur. Arch. Psychiatry Clin. Neurosci. 254:201–208.

Rezaie, P. & Dean, A. (2002). Periventricular leukomalacia, inflammation and white matter lesions within the developing nervous system. Neuropathol. 22:106–132.

Rezaul, I., Persaud, R., Takei, N., Treasure, J. (1996). Season of birth and eating disorders. Int. J. Eat. Disord. 19:53–61.

Rice, F., Jones, I., Thapar, A. (2007). The impact of gestational stress and prenatal growth on emotional problems in offspring: a review. Acta Psychiatr. Scand. 115:171–183.

Richardson, G. A., Goldschmidt, L., Willford, J. (2008). The effects of prenatal cocaine use on infant development. Neurotoxicol. Teratol. 30:96–106.

Richardson, K. (1998). *The Origins of Human Potential.* London: Routledge.

Richardson, K. & Norgate, S. (2005). The equal environment assumption of classical twin studies may not hold. Br. J. Educ. Psychol. 75:339–350.

———. (2006). A critical analysis of IQ studies of adopted children. Hum. Dev. 49:319–335.

Ritvo, E. R., Freeman, B. J., Mason-Brothers, A., Mo, A., Ritvo, A. M. (1985). Concordance for the syndrome of autism in 40 pairs of afflicted twins. Am. J. Psychiatry. 142:74–77.

Ritz, B. & Wilhelm, M. (2008). Ambient air pollution and adverse birth outcomes: methodologic issues in an emerging field. Basic Clin. Pharmacol. Toxicol. 102:182–190.

Robert, J. S. (2000). Schizophrenia epigenesis? Theor. Med. Bioethics 21:191–215.

Roberts, E. M., English, P. B., Grether, J. K., Windham, G. C., Somberg, L., Wolff, C. (2007). Maternal residence near agricultural pesticide applications and autism spectrum disorders among children in the California Central Valley. Environ. Health Perspect. 115:1482–1489.

Robinson, S. (2005). Systemic prenatal insults disrupt telencephalon development: implications for potential intervention. Epilepsy & Behavior. 7:345–363.

Rogers, L. (2001). *Sexing the Brain.* New York: Columbia University Press.

Ronalds, G. A., De Stavola, B. L., Leon, D. A. (2005). The cognitive cost of being a twin: evidence from comparisons within families in the Aberdeen children of the 1950s cohort study. Br. Med. J. 331:1306–1310.

Roohi, J., Montagna, C., Tegay, D. H., Palmer, L. E., DeVincent, C., Pomeroy, J. C., Christian, S. L., Nowak, N., Hatchwell, E. (2009). Disruption of Contactin 4 in 3 Subjects with Autism Spectrum Disorder. J. Med. Genet.46:176–182.

Rorabaugh, W. J. (2003). Drinking in the "Thin Man" films, 1934–1947. Social Hist. Alcohol & Drugs. 18:51–68.

Rosen, D., Seng, J. S., Tolman, R. M., Mallinger, G. (2007). Intimate partner violence, depression, and posttraumatic stress disorder as additional predictors of low birth weight infants among low-income mothers. J. Interpers. Violence. 22:1305–1314.

Rosen, M. G. & Hobel, C. J. (1986). Prenatal and perinatal factors associated with brain disorders. Obstet. Gynecol. 68:416–421.

Ruediger, T. & Bolz, J. (2007). Neurotransmitters and the development of neural circuits. Adv. Exp. Med. Biol. 621:104–115.

Russell, E. W. (2007). Commentary: the Flynn effect revisited. Appl. Neuropsychol. 14:262–266.

Rutter, M. (2006). *Genes and Behavior: Nature-Nurture Interplay Explained.* Malden, MA: Blackwell Publishing.

Sabatini, B. L. (2007). Neighbourly synapses. Nature. 450:1173–1175.

Sadler, T. W. (2006). *Medical Embryology.* 10th edition. New York: Lippincott Williams & Wilkins.

Sadock, B. J. & Sadock, V. A. (2003). *Synopsis of Psychiatry.* 9th edition. New York: Lippincott Williams & Wilkins.

Sampson, P. D., Streissguth, A. P., Bookstein, F. L., Little, R. E., Clarren, S. K., Dehaene, P., Hanson, J. W., Graham, J. M. Jr. (1997). Incidence of fetal alcohol syndrome and prevalence of alcohol-related neurodevelopmental disorder. Teratology. 56:317–326.

Samuels, B. A. & Tsai, L-H. (2004). Nucleokinesis illuminated. Nat. Neuros. 7:1169–1170.

Sarkar, P., Bergman, K., Fisk, N. M., O'Connor, T. G., Glover, V. (2007). Ontogeny of foetal exposure to maternal cortisol using midtrimester amniotic fluid as a biomarker. Clin. Endocrinol. (Oxf). 66:636–640.

Sarkar, P., Bergman, K., O'Connor, T. G., Glover, V. (2008). Maternal antenatal anxiety and amniotic fluid cortisol and testosterone: possible implications for foetal programming. J. Neuroendocrinol. 20:489–496.

Satterfield, J. H., Faller, K. J., Crinella, F. M., Schell, A. M., Swanson, J. M., Homer, L. D. (2007). A 30-year prospective follow-up study of hyperactive boys with conduct problems: adult criminality. J. Am. Acad. Child Adolesc. Psychiatry. 46:601–610.

Savage, C. L., Anthony, J., Lee, R., Kappesser, M. L., Rose, B. (2007). The culture of pregnancy and infant care in African American women: an ethnographic study. J. Transcult. Nurs. 18:215–223.

Savitz, D. A., Chan, R. L., Herring, A. H., Howards, P. P., Hartmann, K. E. (2008). Caffeine and miscarriage risk. Epidemiol. 19:55–62.

Schanen, N. C. (2006). Epigenetics of autism spectrum disorders. Hum. Mol. Genet. 15 Spec. 2:R138–150.

Schechter, R. & Grether, J. K. (2008). Continuing increases in autism reported in California's Developmental Services System. Arch. Gen. Psychiatry. 65:19–24.

Schell, L. M., Gallo, M. V., Denham, M., Ravenscroft, J., (2006). Effects of pollution on human growth and development: an introduction. J. Physiol. Anthropol. 25:103–112.

Scher, M. S. (2003). Prenatal contributions to epilepsy: lessons from the bedside. Epileptic Disord. 5:77–91.

Schettler, T. (2001). Toxic threats to neurologic development of children. Environ Health Perspect. 109(Suppl. 6):813–6.

Schmidt-Kastner, R., Van Os, J., Steinbusch, H., Schmitz, C. (2006). Gene regulation by hypoxia and the neurodevelopmental origin of schizophrenia. Schizophr. Res. 84:253–271.

Schoenemann, P. T., Budinger, T. F., Sarich, V. M., Wang, W. S. (2000). Brain size does not predict general cognitive ability within families. Proc. Natl. Acad. Sci. U. S. A. 97:4932–4937.

Schoenwolf, G. C., Bleyl, S. B., Brauer, P. R., Francis-West, P. H. (2009). *Larsen's Human Embryology*, 4th Edition. Philadelphia: Churchill Livingstone.

Schonfeld, A. M., Mattson, S. N., Riley, E. P. (2005). Moral maturity and delinquency after prenatal alcohol exposure. J. Stud. Alcohol. 66:545–554.

Schuetze, P., Eiden, R. D., Coles, C. D. (2007). Prenatal cocaine and other substance exposure: effects on infant autonomic regulation at 7 months of age. Dev. Psychobiol. 40:276–289.

Seckl, J. R. (2008). Glucocorticoids, developmental 'programming' and the risk of affective dysfunction. Prog. Brain Res. 167:17–34.

Seckl, J. R. & Meaney, M. J. (2006). Glucocorticoid "programming" and PTSD risk. Ann. N. Y. Acad. Sci. 1071:351–378.

Segal, N. L. (2006). Two monozygotic twin pairs discordant for female-to-male transsexualism. Arch. Sex. Behav. 35:347–358.

Shaw, P. (2007). Intelligence and the developing human brain. BioEssays. 29:962–973.

Shaw, P., Eckstrand, K., Sharp, W., Blumenthal, J., Lerch, J. P., Greenstein, D., Clasen, L., Evans, A., Giedd, J., Rapoport, J. L. (2007). Attention-deficit/hyperactivity disorder is characterized by a delay in cortical maturation. Proc. Natl. Acad. Sci. U. S. A. 104:19649–19654.

Shaywitz, S. E., Morris, R., Shaywitz, B. A. (2008). The Education of Dyslexic Children from Childhood to Young Adulthood. Annu. Rev. Psychol. 59:451–475.

Sherrington, C. S. (1937). Man on his nature. Gifford Lectures 1937–1938. In: Chalmers, N., Crawley, R., Rose, S. (eds.) (1971). *The Biological Bases of Behavior.* London: The Open University Press, pp. 289–297.

Siddiqui, A. R., Gold, E. G., Yang, X., Lee, K., Brown, K. H., Bhutta, Z. A. (2008). Prenatal Exposure to Wood Fuel Smoke and Low Birth Weight. Environ. Health Perspect. 116:543–549.

Silvert, M.(2000). Claim that events before birth cause cerebral palsy is disputed. British Med. J. 320:1626.

Singh, G. K. & Kposowa, A. J. (1994). A comparative analysis of infant mortality in major Ohio cities: significance of socio-biological factors. Appl. Behav. Sci. Rev. 2:77–94.

Singh, S. M., Murphy, B., O'Reilly, R. (2002). Epigenetic contributors to the discordance of monozygotic twins. Clin. Genet. 62:97–103.

Siok, W. T., Niu, Z., Jin, Z., Perfetti, C. A., Tan, L. H. (2008). A structural-functional basis for dyslexia in the cortex of Chinese readers. Proc. Natl. Acad. Sci. U. S. A. 105:5561–5566.

Skipper, M. (2008). Not-so-identical twins. Nature. Rev. Genet. 9:250–251.

Skounti, M., Philalithis, A., Galanakis E. (2007). Variations in prevalence of attention deficit hyperactivity disorder worldwide. Eur. J. Pediatr. 166:117–123.

Sokol, R. J., Delaney-Black, V., Nordstrom, B. (2003). Fetal alcohol spectrum disorder. J. Am. Med. Assoc. 290:2996–2999.

Sood, B., Delaney-Black, V., Covington, C., Nordstrom-Klee, B., Ager, J., Templin, T., Janisse, J., Martier, S., Sokol, R. J. (2001). Prenatal alcohol exposure and childhood behavior at age 6 to 7 years: I. dose-response effect. Pediatrics. 108(2):E34.

Srám, R. J., Binková, B., Dejmek, J., Bobak, M. (2005). Ambient air pollution and pregnancy outcomes: a review of the literature. Environ. Health Perspect. 113:375–382.

Steptoe, P. C. & Edwards, R. G. (1978). Birth after the reimplantation of a human embryo. Lancet. Aug 12. 2:366.

Stevens, B., Allen, N. J., Vazquez, L. E., Howell, G. R., Christopherson, K. S., Nouri, N., Micheva, K. D., Mehalow, A. K., Huberman, A. D., Stafford, B., Sher, A., Litke, A. M., Lambris, J. D., Smith, S. J., John, S. W., Barres, B. A. (2007). The classical complement cascade mediates CNS synapse elimination. Cell. 131:1164–1178.

Stevens, J. R. (2002). Schizophrenia: reproductive hormones and the brain. Am. J. Psychiatry. 159:713–719.

Stewart, P., Reihman, J., Gump, B., Lonky, E., Darvill, T., Pagano, J. (2005). Response inhibition at 8 and 9 1/2 years of age in children prenatally exposed to PCBs. Neurotoxicol Teratol. 27:771–80.

Stewart, P. W., Lonky, E., Reihman, J., Pagano, J., Gump, B. B., Darvill, T. (2008). The relationship between prenatal PCB exposure and intelligence (IQ) in 9-year-old children. Environ. Health Perspect.116:1416–1422.

Stotts, A. L., Shipley, S. L., Schmitz, J. M., Sayre, S. L., Grabowski, J. (2003). Tobacco, alcohol and caffeine use in a low-income, pregnant population. J. Obstet. Gynaecol. 23:247–251.

Streissguth, A. P., Barr, H. M., Sampson, P. D. (1990). Moderate prenatal alcohol exposure: effects on child IQ and learning problems at age 7 1/2 years. Alcohol Clin. Exp. Res. 14:662–669.

Stretesky, P. B. & Lynch, M. J. (2004). The relationship between lead and crime. J. Health Soc. Behav. 45:214–229.

Sullivan, K. M. (2008). The interaction of agricultural pesticides and marginal iodine nutrition status as a cause of autism spectrum disorders. Environ. Health Perspect. 116:A155.

Sullivan, P. F., Kendler, K. S., Neale, M. C. (2003). Schizophrenia as a complex trait: evidence from a meta-analysis of twin studies. Arch. Gen. Psychiatry. 60:1187–1192.

Sulloway, F. J. (2007). How to inherit IQ: the fetal question. The New York Review of Books. 54(16), October 25.

Sun, Y., Vestergaard, M., Jakob Christensen, Andre J. Nahmias, Olsen, J. (2008). Prenatal exposure to maternal infections and epilepsy in childhood: a population-based cohort study. Pediatrics. 121:e1100–e1107.

Susser, E., Neugebauer, R., Hoek, H. W., Brown, A. S., Lin, S., Labovitz, D., Gorman, J. M. (1996). Schizophrenia after prenatal famine. Further evidence. Arch. Gen. Psychiatry. 53:25–31.

Suzuki, Y., Toribe, Y., Mogami, Y., Yanagihara, K., Nishikawa, M. (2008). Epilepsy in patients with congenital cytomegalovirus infection. Brain Dev. 30:420–424

Swan, S. H., Waller, K., Hopkins, B., Windham, G., Fenster, L., Schaefer, C., Neutra, R. R. (1998). A prospective study of spontaneous abortion: relation to amount and source of drinking water consumed in early pregnancy. Epidemiology. 9:126–133.

Szasz, T. (1973). The Second Sin. Garden City, NY: Anchor/Doubleday.

Talge, N. M., Neal, C., Glover, V., Early Stress, Translational Research and Prevention Science Network: Fetal and Neonatal Experience on Child and Adolescent Mental Health. (2007). Antenatal maternal stress and long-term effects on child neurodevelopment: how and why? J. Child Psychol. Psychiatry. 48:245–261.

Tamura, T. & Picciano, M. F. (2006). Folate and human reproduction. Am. J. Clin. Nutr. 83:993–1016.

Temkin, O. (1945). The Falling Sickness: A History of Epilepsy from the Greeks to the Beginnings of Modern Neurology. Baltimore: The Johns Hopkins University Press.

Testa, M., Quigley, B. M., Eiden, R. D. (2003). The effects of prenatal alcohol exposure on infant mental development: a meta-analytical review. Alcohol & Alcoholism. 38:295–304.

The White House. (2007). President Bush visits Cleveland, Ohio. July 10, 2007. http://www.whitehouse.gov/news/releases/2007/07/20070710–6.html (available May 6, 2008).

Tobin, A. J. (1999). Amazing grace: sources of phenotypic variation in genetic boosterism. In: Carson, R. A. & Rothstein, M. A. (eds.) *Behavioral Genetics: The Clash of Culture and Biology.* Baltimore: The Johns Hopkins University Press, pp. 1–11.

Tomkins, A., Murray, S., Rondo, P., Filteau, S. (1994). Impact of maternal infection on foetal growth and nutrition. SCN News. 11:18–20.

Tong, S., Von Schirnding, Y. E., Prapamontol, T. (2000). Environmental lead exposure: a public health problem of global dimensions. Bull. World Health Organ. 78:1068–1077.

Toro, R., Leonard, G., Lerner, J. V., Lerner, R. M., Perron, M., Pike, G. B., Richer, L., Veillette, S., Pausova, Z., Paus, T. (2008). Prenatal exposure to maternal cigarette smoking and the adolescent cerebral cortex. Neuropsychopharmacol. 33:1019–1027.

Torrey, E. F., Miller, J., Rawlings, R., Yolken, R. H. (1997). Seasonality of births in schizophrenia and bipolar disorder: a review of the literature. Schizophr. Res.28:1–38.

Trabert, B, & Misra, D. P. (2007). Risk factors for bacterial vaginosis during pregnancy among African American women. Am. J. Obstet. Gynecol. 197:477.e1–477.e8.

Trasande, L., Landrigan, P. J., Schechter, C. (2005). Public health and economic consequences of methylmercury toxicity in the developing brain. Environ. Health. Perspect. 113:590–596.

Trasande, L, Landrigan, P. J., Schechter, C. B., Bopp, R. F. (2007) Methylmercury and the developing brain. Environ. Health Perspect. 115:A396-A3977; author reply A397-A398.

Trasande, L., Schechter, C. B., Haynes, K. A., Landrigan, P. J. (2006). Mental retardation and prenatal methylmercury toxicity. Am. J. Industr. Med. 49:153–158.

Tsai, J., Floyd, R. L., Bertrand, J. (2007). Tracking binge drinking among U.S. childbearing-age women. Prev. Med. 44:298–302.

Tsai, J., Floyd, R. L., Green, P. P., Boyle, C. A. (2007b). Patterns and average volume of alcohol use among women of childbearing age. Matern. Child Health J. 11:437–445.

Tsuang, M. (2000). Schizophrenia: genes and environment. Biol. Psychiatry. 47:210–220.

Turkheimer, E., Haley, A., Waldron, M., D'Onofrio, B., Gottesman, I. I. (2003). Socioeconomic status modifies heritability of IQ in young children. Psychol. Sci. 14:623–28.

Tustin, K., Gross, J., Hayne, H. (2004). Maternal exposure to first-trimester sunshine is associated with increased birth weight in human infants. Dev. Psychobiol. 45:221–230.

Ui, J. (1992). Minamata disease. In: Ui, J. (ed.) *Industrial Pollution in Japan.* Chapter 4. Tokyo: United Nations University Press. http://www.unu.edu/unupress/unupbooks/uu35ie/uu35ie0c. htm#chapter%20%20%204%20minamata%20disease

United States Department of Health and Human Services. Surgeon General. News release, February 21, 2005. U.S. Surgeon General Releases Advisory on Alcohol Use in Pregnancy. http://www. hhs.gov/surgeongeneral/pressreleases/sg02222005.html

Urakubo, A., Jarskog, L. F., Lieberman, J. A., Gilmore, J. H. (2001). Prenatal exposure to maternal infection alters cytokine expression in the placenta, amniotic fluid, and fetal brain. Schizophr. Res. 47:27–36.

U.S. Preventive Services Task Force. (2006). Screening for Speech and Language Delay in Preschool Children: Recommendation Statement. Originally in Pediatrics 2006 117:497–501. February 2006. Agency for Healthcare Research and Quality, Rockville, MD. http://www.ahrq.gov/clinic/uspstf06/speech/speechrs. htm

Van der Zee, H. A. (1982). *The Hunger Winter*. London: J. Norman & Hobhouse.

Van Erp, T. G., Saleh, P. A., Rosso, I. M., Huttunen, M., Lönnqvist, J., Pirkola, T., Salonen, O., Valanne, L., Poutanen, V. P., Standertskjöld-Nordenstam, C. G.,Cannon, T. D. (2002). Contributions of genetic risk and fetal hypoxia to hippocampal volume in patients with schizophrenia or schizoaffective disorder, their unaffected siblings, and healthy unrelated volunteers. Am. J. Psychiatry. 159:1514–1520.

Van Os, J. & Selten, J. P. (1998). Prenatal exposure to maternal stress and subsequent schizophrenia. The May 1940 invasion of The Netherlands. Br. J. Psychiatry. 172:324–326.

Ventolini, G. & Neiger, R. (2006). Placental dysfunction: pathophysiology and clinical considerations. J. Obstetr. Gynecol. 26:728–730.

Vreugdenhil, M. J., Lanting, C. I., Mulder, P. G., Boersma, E. R., Weisglas-Kuperus, N. (2002). Effects of prenatal PCB and dioxin background exposure on cognitive and motor abilities in Dutch children at school age. J. Pediatr. 140:48–56.

Wachs, T. D., Pollitt, E., Cueto, S., Jacoby, E., Creed-Kanashiro, H. (2005). Relation of neonatal iron status to individual variability in neonatal temperament. Dev. Psychobiol. 46:141–153.

Wadhwa, P. D. (2005). Psychoneuroendocrine processes in human pregnancy influence fetal development and health. Psychoneuroendocrinol. 30:724–743.

Waller, K., Swan, S. H., DeLorenze, G., Hopkins, B. (1998). Trihalomethanes in drinking water and spontaneous abortion. Epidemiology. 9:134–140.

Wang, H-L., Chen, X-T., Yang, B., Ma, F. L., Wang, S., Tang, M. L., Hao, M. G., Ruan, D. Y. (2008). Case-Control Study of Blood Lead Levels and Attention-Deficit Hyperactivity Disorder in Chinese Children. Environ. Health Perspect. 116:1401–1406. Wang, L. & Pinkerton, K. E. (2007). Air pollutant effects on fetal and early postnatal development. Birth Defects Res. C Embryo Today. 81:144–154.

Watson, J. B., Mednick, S. A., Huttunen, M., Wang, X. (1999). Prenatal teratogens and the development of adult mental illness. Dev. Psychopathol. 11:457–466.

Watson, P. E. & McDonald, B. W. (2007). Seasonal variation of nutrient intake in pregnancy: effects on infant measures and possible influence on diseases related to season of birth. Eur. J. Clin. Nutr. 61:1271–1280.

Watson, S. M. & Wesby, C. E. (2003). Strategies for addressing the executive function impairment of students prenatally exposed to alcohol and other drugs. Communication Disord. Q. 24: 194–204.

Wattendorf, D. J. & Muenke, M. (2005). Fetal alcohol spectrum disorders. Am. Family Physician. 72:279–285.

Wei, J. & Hemmings, G. P. (2005). Gene, gut and schizophrenia: the meeting point for the gene-environment interaction in developing schizophrenia. Med. Hypotheses. 64:547–552.

Weiss, B. & Bellinger, D. C. (2006). Social ecology of children's vulnerability to environmental pollutants. Environ. Health Perspect. 114:1479–1485.

Werner, E. A., Myers, M. M., Fifer, W. P., Cheng, B., Fang, Y., Allen, R., Monk, C. (2007). Prenatal predictors of infant temperament. Dev. Psychobiol. 49:474–484.

Whitehead, N. S. & Leiker, R. (2007). Case management protocol and declining blood lead concentrations among children. Prev. Chronic Dis. (serial online). Volume 4: No. 1, January 2007. 2007http://www.cdc.gov/pcd/issues/2007/jan/06_0023.htm

Whittle, N., Sartori, S. B., Dierssen, M., Lubec, G., Singewald, N. (2007). Fetal Down syndrome brains exhibit aberrant levels of neurotransmitters critical for normal brain development. Pediatrics. 120:e1465–1471.

Whyatt R. M., Barr, D. B., Camann, D. E., Kinney, P. L., Barr, J. R., Andrews, H. F., Hoepner, L. A., Garfinkel, R., Hazi, Y., Reyes, A., Ramirez, J., Cosme, Y., Perera, F. P. (2003). Contemporary-use pesticides in personal air samples during pregnancy and blood samples at delivery among urban minority mothers and newborns. Environ. Health Perspect. 111:749–756.

Whyatt, R. M., Camann, D. E., Kinney, P. L., Reyes, A., Ramirez, J., Dietrich, J., Diaz, D., Holmes, D., Perera, F. P. (2002). Residential pesticide use during pregnancy among a cohort of urban minority women. Environ. Health Perspect. 110:507–514.

Whyatt, R. M., Rauh, V., Barr, D. B., Camann, D. E., Andrews, H. F., Garfinkel, R., Hoepner, L. A., Diaz, D., Dietrich, J., Reyes, A., Tang, D., Kinney, P. L., Perera, F. P. (2004). Prenatal insecticide exposures and birth weight and length among an urban minority cohort. Environ. Health Perspect. 112:1125–1132.

Willford, J., Leech, S., Day, N. (2006). Moderate prenatal alcohol exposure and cognitive status of children at age 10. Alcohol Clin. Exp. Res. 30:1051–1059.

Wilson, C. A. & Davies, D. C. (2007). The control of sexual differentiation of the reproductive system and brain. Reproduction. 133:331–359.

Wilson, D. S. & Wilson, E. O. (2007). Rethinking the theoretical foundation of sociobiology. Q. Rev. Biol. 82:327–348.

Wilson, E. O. (1975). *Sociobiology: The New Synthesis.* (25th anniversary ed., 2000). Cambridge, MA.: Harvard University Press.

Windham, G. & Fenster, L. (2008). Environmental contaminants and pregnancy outcomes. Fertil. Steril. (89 Suppl. 1):e111–116.

Wolff, M. S., Teitelbaum, S. L., Lioy, P. J., Santella, R. M., Wang, R. Y., Jones, R. L., Caldwell, K. L., Sjödin, A., Turner, W. E., Li, W., Georgopoulos, P., Berkowitz, G. S. (2005). Exposures among pregnant women near the World Trade Center site on 11 September 2001. Environ. Health Perspect. 113:739–748.

World Health Organization. (2007a). Newborns with low birthweight. http://www.who.int/whosis/database/core/

World Health Organization. (2007b). Infant mortality rate. http://www.who.int/whosis/database/core/

Wright, J. P., Dietrich, K. N., Ris, M. D., Hornung, R. W., Wessel, S. D., Lanphear, B. P., Ho, M., Rae, M. N. (2008). Association of prenatal and childhood blood lead concentrations with criminal arrests in early adulthood. PLoS Med. 5(5):e101-e109.

Yeargin-Allsopp, M., Rice, C., Karapurkar, T., Doernberg, N., Boyle, C., Murphy, C. (2003). Prevalence of autism in a US metropolitan area. J. Am. Med. Assoc. 289:49–55.

Yehuda, R., Engel, S. M., Brand, S. R., Seckl, J., Marcus, S. M., Berkowitz, G. S. (2005). Transgenerational effects of posttraumatic stress disorder in babies of mothers exposed to the World Trade Center attacks during pregnancy. J. Clin. Endocrinol. Metab. 90:4115–4118.

Yelin, R., Kot, H., Yelin, D., Fainsod, A. (2007). Early molecular effects of ethanol during vertebrate embryogenesis. Differentiation. 75:393–403.

Yelin, R., Schyr, R. B., Kot, H., Zins, S., Frumkin, A., Pillemer, G., Fainsod, A. (2005). Ethanol exposure affects gene expression in the embryonic organizer and reduces retinoic acid levels. Dev. Biol. 279:193–204.

Yoon, B. H., Park, C. W., Chaiworapongsa, T. (2003). Intrauterine infection and the development of cerebral palsy. Br. J. Obstetr. Gynecol. (110 Suppl 20):124–127.

Yuan, W., Holland, S. K., Cecil, K. M., Dietrich, K. N., Wessel, S. D., Altaye, M., Hornung, R. W., Ris, M. D., Egelhoff, J. C., Lanphear, B. P. (2006). The impact of early childhood lead exposure on brain organization: a functional magnetic resonance imaging study of language function. Pediatrics. 118:971–977.

Zammit, S., Lewis, S., Gunnell, D., Smith, G. D. (2007). Schizophrenia and neural tube defects: comparisons from an epidemiological perspective. Schizophr. Bull. 33:853–8588.

Zhao, X., Pak, C., Smrt, R. D., Jin, P. (2007). Epigenetics and neural developmental disorders: Washington DC, September 18 and 19, 2006. Epigenetics. 2:126–34.

Zhou, J-N., Hofman, M. A., Gooren, L. J., Swaab, D. F. (1995). A sex difference in the human brain and its relation to transsexuality. Nature. 378:68–70.

Zierold, K. M. & Anderson, H. (2004). Trends in blood lead levels among children enrolled in the special supplemental nutrition program for women, infants, and children from 1996 to 2000. Am. J. Public Health. 94:1513–1515.

INDEX

ABC News, 193
"Ablatio penis", 158. *See also* Penile ablation
Acetaldehyde, 109–110
Acetone, 109
Adolescence/adolescents, 37, 53, 102, 156,
    192, 197, 211–212, 229, 276. *See also*
    Adolescent pregnancy
Adolescent mothers. *See* Adolescent pregnancy
Adolescent pregnancy, 276, 294–295
Adoption studies, 253–254
Adrenal glands, 6, 51–52, 281–283
Adult behavior. *See* Childhood and adult
    behavior
Adult brain, 240–241
Adult hand morphology, 163
Adult intelligence, 14–15, 17, 233–234, 246, 257
Adult life, 81–82, 87–89, 121–122
Adult offspring, 285. *See also* Offspring
Adult pathology, 221–222, 223–226
Adult phenotype, 63–64, 100
Adult schizophrenia, 220–222. *See also*
    Schizophrenia
Adulthood, 7, 53, 81, 88, 211, 212, 286

Affective disorders/affective disorders, 212,
    219, 231, 285
Affluent families, 14–15, 255–256
African-American children, 137
African-American women, 118, 293
After birth events, 45–46, 150, 160, 184, 223,
    247, 265
    autistic symptoms, 201
    brain development, 54, 73, 183, 186
    Darwinian programming, 145–146
    neurological and behavioral deficits, 96
Agassiz, Louis, 10
Air pollution/air pollutants, 119–122
Alcohol consumption/alcoholism, 13, 22,
    293, 295. *See also* Pregnant mothers;
    Pregnant women
    abuse, 132, 139
    fetal/prenatal impact of, 130–141, 242–243
    long history of, 127–129
    popular beliefs about, 263–264
Alcoholic beverages, 103, 128, 140
Alcohol-related neurodevelopmental disorder
    (ARND), 131

Aldosterone, 165
Alzheimer's disease, 88
American black children, 28
American children, 28–30, 170
    ADHD in, 186
    ethyl alcohol exposure, 127
    neuropsychological developmental
        disabilities, 170
    umbilical cord mercury blood levels of, 114
American Indians, 136
American medical community/medical
    community, 138–139, 205,
        208–209, 247
American non-Hispanic white children, 28
American psychiatry, 166, 191–193
American-Hispanic women, 277
Americans, 3–7, 10, 25, 116, 128, 234–235
    epilepsy incidence, 177
    iodine nutrition status of, 200
    neuropsychological developmental
        disabilities in, 170
    and schizophrenia, 211–212
Amnion/amniotic membrane, 194, 246, 277
Amniotic cavity, 101
Amniotic fluid, 6, 99, 101, 134, 246
Anabolic steroids, 152
Analgesics, 103
Anatomy, 81, 149, 162, 178–179, 180, 197, 276,
    281
Androgens/androgenic hormones, 119, 162–165
Anencephaly, 100
Animal behavior, 19, 80
Animal experiments, 80, 92, 116, 139–140, 286
Animal oddities, 146–150
Anorexia nervosa, 210, 229–230, 231
Antibacterials, 103
Antibodies, 166, 178
Anticoagulants, 103
Anticonvulsants, 102
Anti-lead treatment, 27
Antisocial behavior. See Social behavior/
    antisocial behavior
Anxiety disorders, 15, 208, 224–225
Apoptosis, 133
Appalachia, 289
Argentina, 257
Aristotle, 149
Arsenic, 22, 53, 115

Artificial sex determination, 151
Asian-Indian-American women, 268
Asperger syndrome, 189
Atlanta, Georgia, 171
Attention, 40
Attention deficit hyperactivity disorder
    (ADHD), 17, 21, 170–171, 185–187,
        194, 208, 300
    lead exposure, 38–39
    stressful events in pregnancy, 285–287
Attention deficits. See Attention deficit
    hyperactivity disorder (ADHD)
Australia, 28, 30, 229, 257, 291, 300
Austria, 191, 257
Autism, 17, 25–27, 170–171, 204–205, 218, 267
    correlations, 202–203
    diagnostic categories of, 187, 190–193
    epigenetic etiology of, 198
    and identical twins, 194
    likely causes of, 199–202
    neurobiology of, 196
    symptoms of, 27, 201
Autism spectrum disorder (ASD), 189–190,
    191, 199, 201, 205. See also "Broad"
    autism (autism spectrum disorder,
    ASD)
Autistic behavior, 187–189, 191–193, 196–198,
    205
Autistic children, 26–27, 192–193, 197,
    204–205
    non-Hispanic mothers of, 200
    schizophrenia incidence in, 199
Automobile behavioral trait, 72
Autonomic nervous system, 78, 224, 227
Awareness, 176, 227
Axons, 80, 100
Ayurvedic medicines, 34

Barker, David, 87
Barker hypothesis, 87–88
Bay of Naples, Italy, 146
Bayley Scales of Infant Development, 31, 138
Before birth events, 79, 184, 196
Behavior geneticists, 236, 238, 242, 243, 249
Behavior genetics/behavioral genetics,
    202–204, 235–240, 249
Behavior patterns, 14, 82, 88, 166, 298
Behavior/behavioral traits, 72, 130, 160, 254

Behavioral consequences, 37, 121–122
Behavioral deficits, 36, 96–97, 139
Behavioral dysfunction, 17, 95, 183
Behavioral effects, 14, 37, 39–41, 137–141
Behavioral geneticists, 183–184, 216–218, 239, 313n.8
Behavioral phenotype, 195–196
Behavioral problems/difficulties, 15, 38, 120, 134, 137, 169, 171, 271, 277
Belgium, 257
Bettelheim, Bruno, 191–193
"Individual and Mass Behavior in Extreme Situations" (paper), 191
Binge drinking, 126, 133, 135–136, 138, 230, 263–264
Biochemical cascade, 103
Biochemical entities, 56
Biochemical events/changes, 80, 201
Biochemical processes, 50
Biochemical reactions, 63, 67, 104
Biochemical system, 37, 39
Biochemical-reaction routines/paths, 63, 67, 104
Biogenic mechanisms, 291
Biological evolution, 11
Biological identity, 148–149
Biological molecules, 103
Biological psychiatry, 78
Biological sex, 151
Biological systems, 50, 93–94, 104–105, 109–110
Biologists, 10, 11, 12, 44, 61, 63–64, 150, 215, 238, 239
Bipolar disorder, 209, 212–214, 218–232
Birth asphyxia, 174
Birth cohort 1800–1870, 97
Birth defects, 20–21, 25–26, 134, 170, 171. See also Congenital defects/anomalies
Birth outcomes, 132, 245
Birth weight, 13–14, 97, 268, 271, 272, 276, 286. See also Low-birth-weight children
Bisexuality, 148–150, 157–158, 166–167
Black children, 28
Blacks, 28, 136, 200, 267, 268, 274, 275, 289, 295
Black–white differences/gaps/disparities, 267–268, 274–275
Blastocysts, 46, 58–59, 106, 246. See also Morula
Blindness, 112–115, 170. See also Childhood blindness

Blood concentration, 105–107. See also Blood lead concentrations; Blood lead level (BLL) standards
Blood lead concentrations, 30–38
Blood lead level (BLL) standards, 27–32, 242, 300
Blood pressure, 224, 282, 286. See also Hypertension/postnatal hypertension
Blood serum, 38–39
Blood sugar level, 282
Body segments, 61–62
Bogle family, the, 297
Bone, 20, 31, 34, 107, 296
Bonellia (marine worms), 146
Borrelia burgdorferi, 201
Boston, 28
Bovie cautery device, 154
Brain cells, 49, 133
Brain damage, 30, 35, 100, 101, 130, 134–135, 139, 172, 174, 177, 222
destruction of cerebral white matter, 172
Brain development/fetal brain development, 17, 70, 102–103, 215, 245, 266, 284. See also Identical twins; Monozygotic (MZ) twins
cascade of processes, 67–68
environmental factors in, 119, 243
lead poisoning and, 35
maternal behavior and, 278–279
maternal infection and, 220–222
neurodegenerative diseases, 86–87
prenatal exposure to
alcohol, 140–141
PCBs, 117–118
processes of, 73–84
subclinical changes in, 100–101
Brain function, 79–81, 137–138, 286–287
Brain gender, 148–150, 160–162
Brain injury/fetal brain injury, 96, 138, 174–176
Brain lesions (periventricular leukomalacia), 172–173
Brain plasticity, 54, 68–79
Brain volume, 240–241
Brain wiring, 70–73, 81, 140. See also Hard-wiring
Brazil, 257
Breast/breast development, 144, 156, 159, 279

British Royal Society, 8
"Broad" autism (autism spectrum disorder, ASD), 189
Bronx Veterans Affairs Medical Center, 4
Bryan, William Jennings, 10
Bulimia nervosa, 229, 230, 231
Burt, Cyril, 237
Bush, President George W., 114
Butanol, 109

Caffeine. See Caffeine consumption/coffee consumption
Caffeine consumption/coffee consumption, 103, 265–266, 278
Canada, 257, 300
Cancer, 68, 116, 119, 282
Carbohydrates, 56
Carboxyhemoglobin, 101
Carcinogens, 116, 119–120
Cardiac abnormalities, 100
Cardiac defects, 20, 126, 131
Cardiovascular disease/defects, 121, 131, 224, 286
Cardiovascular drugs, 103
Cascades, 9, 55–59, 75, 103–104
    description of, 9–10
    of gene-switching, 62–64
    of human brain development, 67–68
Catastrophe/catastrophic events, 4–5, 121, 154, 213. See also Disasters
Caucasian women, 293
Caudate nucleus, 139
Celiac disease, 221–222
Cell differentiation/cellular differentiation, 67, 74, 90–91, 283–284
Cell migration/nerve cell migration, 47, 76, 78, 83
Cell organelles, 56
Cell proliferation/cellular proliferation, 57
Cellular pathology, 177
Cellular toxicology, 186–187
Cellulose acetate, 109
Central nervous system, 7, 220, 224
Cerebral cortex/fetal cerebral cortex, 37, 78, 82–84, 96, 100, 102, 113, 140–141
    of adolescent autistic children, 197
    general architecture of, 71–72
    of human brain, 79, 89

nervous system of, 73–74
neural circuits in, 76–77
neuronal connections in, 68–70, 241
Cerebral ischemia, 174
Cerebral palsy, 112–115, 169, 171–177, 267
Chelation therapy/chelating agents, 27, 34–35
Chemical messengers, 160, 281
Chemical pollutants. See Chemicals/chemical elements
Chemical pollution. See Chemicals/chemical elements
Chemicals/chemical elements, 8, 15–16, 40, 52, 99, 115–118, 175, 243, 278, 289–290
    air pollutants/air pollution, 119–122, 270–272
    cigarette smoke, 101–102
    metal poisoning, 108–114
    prenatal vulnerability to, 57–58, 87–88
Chemotherapy, 35
Chernobyl radiation disaster, 284
Chicago, Illinois, 29, 171, 265, 289
Child behavior, 133, 291. See also Autism; Autistic children
Child-bearing age, 140–141, 279–280
Childhood, 6, 15, 17, 53–55, 79, 86, 88, 120–121, 145–146, 189, 192, 199, 214, 221–222, 224, 232, 270–271, 286, 306. See also Childhood and adult behavior
    blindness/hearing loss in, 179–180
    brain wiring/development in, 70–72, 81–84, 117–118, 197, 269, 290
    and gender identity, 155–160
    and intelligence quotient (IQ), 233, 241–243
    lead poisoning, 35–36, 37–39, 186
    learning disabilities, 181–183
    psychological development, 192–193
    temperaments, 226–227
    and violence, 299–300
Childhood and adult behavior, 13–17, 61–64, 145, 224, 269, 271, 306–307. See also Childhood
Childhood asthma, 120, 270
Childhood behavior. See Childhood and adult behavior
Childhood blindness, 179–180. See also Blindness
Childhood disintegrative disorder, 189

Childhood lead exposure. *See* Lead exposure/
       prenatal lead exposure
Childhood verbal deficits, 6
Chimpanzee behavior, 19
China, 25, 30, 284
Chorioamnionitis, 175
Chorioamniotic membranes, 101
Chorion, 194, 246–248
Chromatin/chromatin proteins, 50
Chromosomal abnormalities, 91–92
Chromosomal defects, 126
Chromosome 17 (LIS1), 82
Chromosome 46 XY, 157
Chromosomes, 47, 49–50, 72, 91, 147–148,
       161–162, 215
Cigarettes, 22, 101–102, 235, 277, 292–293.
       See also Tobacco smoke/effects
Cincinnati, 28, 272
Circumcision, 154
Classical autism, 27, 138, 170–171, 187–192
Clean Air Act (U.S.), 114
Clear Skies Act (U.S.), 114
Cleveland, 28, 287
Clinical diagnoses, 19–21, 228
Clinical entity, 13, 182, 191, 214, 216
Clinical research, 28, 29
Clinicians, 96, 99, 173, 180, 207–208, 213
Clitoris, 151–153
Cocaine, 22, 39, 102–103, 181, 265
Coffee drinking/consumption. *See*
       Caffeine consumption/coffee
       consumption
Cognitive development, 69–70, 169–172
Cognitive disabilities. *See* Cognitive function
Cognitive effects, 117–118, 127
Cognitive function, 27, 35–39, 137–141
Cognitive outcome, 69–70
Cognitive performance, 19, 88, 138, 237,
       250–254
    brain volume as indicator of, 241
    behavior, 76–77, 84
    childhood, 117, 121
    environmental/prenatal impact, 74,
       241–243, 247, 284, 285
    intellectual function, 181, 233, 238, 275–276
Cognitive phenotypes, 96–97
Collateral damage, 140–141, 294
Colon, 279

Combustion emissions, 119–120
Communities, 18, 29–32, 291–293, 305, 307
    cultural factors, 270–280, 285
    environmental lead poisoning, 29–32, 36
Complexity, 48–49, 57, 59, 177, 184, 221
    of autistic behavior, 191
    clinical, 177
    of human brain, 231–232
Conception, 3, 15, 18, 40, 53, 62, 153, 215,
       265–266
    alcohol consumption and, 13, 135–136
    and sexual differentiation, 145, 161
Conditioned stimulus, 140
Congenital adrenal hyperplasia (CAH), 165
Congenital anomalies. *See* Congenital defects/
       anomalies
Congenital defects. *See* Congenital defects/
       anomalies
Congenital defects/anomalies, 120–122, 127,
       165, 178–179, 195, 220, 245, 270–271,
       274. *See also* Birth defects
Congenital German measles (rubella), 194
Contact dermatitis, 175
Contaminated water, 117, 118
Contamination, 15–16, 115–122, 235–236, 273
    chemical, 109–115
    lead, 25–27, 34–36
Convulsions, 109, 176–177
Coronary heart disease, 282
Corpus-callosum, 139
Correlations, 6, 13, 37, 89, 92, 202–204, 217,
       229, 230, 236, 291, 293
    ASD in children, 199
    eating disorders, 230
    explanation for differences in, 203–204, 218
    fetal environment, 84, 231–232
    heritability of intelligence quotient (IQ),
       216–217
    IQ scores for identical twins, 247–254
    low birth weight, 225, 278, 288
    major depression, 229
    schizophrenia, 15–16, 219
    season of birth, 218–219, 230–231
Cortical malformations, 177
Cortical neurons, 81
Cortical synapses, 68
Cortisol/cortisol levels, 5–7, 51–52, 101, 224,
       286–287

Cortisol/cortisol levels (*cont.*)
late gestation effects, 288
physiological response to stress, 282,
283–284
prenatal cocaine exposure, 102
Cortisol reactivity, 286
Cortisol secretion. *See* Cortisol/cortisol levels
Cotwins, 164–165, 178–179, 246, 251, 252–253.
See also entries for Twins
Crepidula fornicate (slipper-snail), 146
Crick, Francis, 11
Criminality/criminal behavior, 18, 22,
259–260
fetal environment, 295–300
innate, 301–305
Cultural factors/determinants, 11, 18, 22,
96–97, 182–184, 250–251, 279–280
maternal behavior, 263–266, 274–279
"negative eugenics", 234–235
socioeconomic status, 268–273
Cultural influences, 266, 276–280
Cultural pressures, 22, 275–280
Culture, 22, 53, 165, 208–209, 225–226, 257,
270–273, 280–288. *See also* Poverty
and fetal environment, 266
IQ testing and, 235–237, 250
and low birth weight and infant mortality,
267–269
maternal behavior during pregnancy and,
274–279
and maternal stress on the fetus, 280–286
and popular beliefs, 263–266
socially visible forms of, 227–228
Cytokines (hormone-like signaling proteins),
100–101, 175, 220, 283–284
Cytomegalovirus (CMV), 178, 180, 274
Cytosine, 90
Czech Republic, 270

Darwin, Charles, 9, 10, 60, 64, 295, 301
The Descent of Man (1879), 301
On the Origin of Species, 10
Darwinian natural selection, 129–130, 258. *See
also* Natural selection
Darwinian programming, 145–146
Darwin's theory of evolution/Darwinian
evolution, 10, 60, 129, 145, 146, 147
Daughter cells, 90

Deafness, 112–115, 169, 179–180. *See also*
Hearing/hearing loss/hearing
deficits
Deer tick, 201
Degenerative diseases, 97
Delusions, 210–212
Dendrites, 100
Denmark, 177, 257, 278
Dentine lead levels, 171
Deoxyribonucleic acid molecules. *See* DNA
Depression, 6, 101, 199, 212, 213, 225,
229–230, 231
poverty, 291
stressful experiences, 284, 285, 288
Destiny/destinies, 3, 50, 52, 54
Destitute children, 187–188
Determinants, 11, 16, 52, 192–193, 306
of genetic psychiatric disorders, 215
of human behavior, 11, 52–55
of human sexual orientation, 163–167
of intelligence, 243
Detoxification, 126, 258
Developed countries, 30, 271
Developing brain, 7, 29, 35, 57, 78, 96, 107, 108,
241, 283
autoimmune process in, 200
BLL levels and, 36
critical gestational period and, 152
effects of alcohol concentrations on, 58–59,
129, 139
epigenetics and, 76, 94–95
fetal environmental impact on, 299
implications of damage to, 174–176, 179, 290
and sex hormones, 153–154
Development (ontogenesis), 50–52, 58–59
Developmental brain disabilities. *See*
Neuropsychological developmental
disabilities
Developmental cascade. *See* Cascades
Developmental deficits, 27, 102, 188. *See also*
Autism
Developmental disabilities, 17, 169–172,
179–180, 183–184
Developmental genes, 62–64
Developmental origins hypothesis, 87
Developmental origins of disease
hypothesis, 87
Developmental psychobiology, 6–7

Developmentally disabled children, 27
Diabetes, 7, 87, 101, 231, 282
Diagnostic categories, 182, 188–193, 210
Dibenzofurans, 117–118
Dichloro-1. diphenyl-trichloroethane
    (DDT), 185
Dichorionic MZ twins, 248
Dimethylmercury, 110–112
Dioxins, 39, 115–118, 183
Diplegia, 173
Disasters, 4, 12, 281, 284. *See also* Catastrophe/
    catastrophic events
Discordances (in twins), 164, 244, 247
    for behavior/IQ, 244–245
    for dyslexia, 184
    for epilepsy, 178–179
    for schizophrenia, 216–218
    for sexual orientation, 164–167
Diseases, 81, 85, 88, 97, 101, 122, 201, 214–215,
    282, 294, 298
Dizygotic (DZ) twins. *See* Fraternal twins
DNA, 11, 23, 47–50, 52, 54, 96, 106, 120,
    193, 214, 270, 278. *See also* DNA
    methylation/demethylation patterns
    base sequence, 94
    microdeletion, 82–83
    prenatal development, 55
    toxin molecule interaction with, 104
    transcription of, 298
"DNA Age", 11
DNA code, 48, 49–50, 84
DNA methylation/demethylation patterns, 49,
    90–91, 119. *See also* DNA
DNA nucleotide base, 90
DNA sequences, 50, 94
DNA-adducts, 120, 270
Dominica, 257
Dominican women, 290
Dopamine, 198, 215, 227, 229
Dose–response relationship, 104–105
Down syndrome, 82
Drinking during pregnancy, 133–137, 138–141,
    265–266, 293
Drinking patterns, 135
Drunkenness, 132
Dugdale, Richard L., 296
Dutch Hunger Winter/Hunger Winter,
    85–96

Dutch men, 259
Dysfunctional temperaments, children with,
    226–227, 229
Dyslexia, 181–184
Dysmorphologies, 20–21, 131, 137, 204
Dysphoric feelings, 157
Dysthymic (minor depressive) disorder, 228

Early pregnancy, 126, 136
Earth, 11, 45
Earthquakes, 6, 7, 281, 284
Eating disorders, 97, 229–232
Echolalia, 27
Edo/Tokugawa Period (Japan), 108
Edwards, Robert G., 44
Egg cells/female egg cells, 8, 10–11, 44–46
Einstein, Albert, 241
Electrocautery circumcision, 157
Electrosurgical dissection, 154
Embryology/embryologists, 9–10, 59
Embryonic development/human embryonic
    development, 17, 90, 127, 163
Embryonic stem cells, 47
Embryos, 46, 73–74, 90–93, 151–152, 191, 194
    and cascade process of development, 67–68
    fatty acids in nutrition, 99
    and fetuses, 16, 38, 46, 48, 51, 56, 57, 58–59,
      60–62
    gene control of development of, 62–64, 71
    and heterosexuality in males and females, 163
    and MZ twins, 194, 246
    neural development continuity of, 81–82
    vulnerabilities of, 102–103, 104, 105–107,
      126–127, 134–135, 161, 184
Emotional behavior, 80, 223–225
Emotional problems, 169, 171, 192, 225, 284
Emotional refrigerators, 205
Emotional stability, 121
Emotional stress, 26, 51–52
Emotions, 19, 53, 187, 226–227
Encephalitis, 186
Encephalopathy, 200–201
Endocrine disruptors, 118–119, 221
Endocrine effects, 145
Endophenotypes, 95
End-organ vulnerability, 15
English-speaking (acculturated) Hispanic
    women, 135, 295

Environmental agents, 40–41, 242, 265–266
Environmental changes, 77, 258
Environmental conditions, 6, 130
Environmental effects, 3, 46, 59, 67–68, 72,
    75–76, 87, 92, 93, 305
    and Barker hypothesis, 121–122
    on brain development, 67–68, 76
    and identical (monozygotic) twins, 46
Environmental events, 6–7, 50
Environmental factors, 16, 20–21, 217, 272–274
Environmental impact, 47–48, 74, 82, 96,
    121–122, 160–162
    developmental consequences of, 92–93
    fetal, 194, 299
    producing alterations, 88–89
    on tissue development, 90–91
Environmental lead pollution, 25, 27–39. See
    also Lead exposure/prenatal lead
    exposure; Lead poisoning
Environmental pollutants/pollutants, 41, 221
Environmental stressors, 205
Environmental toxins/chemicals, 17, 21–23,
    55–56, 90–91, 180
    and childhood health, 14–15
    prenatal impact of, 171–172, 271
Environment-dependent gene expression, 196
Epidemiological studies, 13, 100, 219
Epidemiologists, 87, 185
Epidemiology, 174, 275
Epigenetic etiology (of autism), 198
"Epigenetic events", 52–53, 89
Epigenetic inheritance,   93–96
Epigenetic modifications, 91
Epigenetic mutations, 198
Epigenetics, 51, 63–64, 89, 93–96, 298
    developing brain, 76
    neurobiology and,   196–199
Epilepsy/epileptic seizures, 176–179
Epileptic births, 177–178
Estonia, 257
Estrogens, 119, 165
Ethanol, 103–104, 126, 129–130, 133, 183, 265. See
    also Ethyl alcohol
Ethnic groups, 35, 136–137, 268, 271–272,
    274–276, 294–295
    autism spectrum disorder (ASD) in, 190
    differences in intelligence between, 235
    maternal diet of, 279–280

Ethnic/cultural heritage, 18, 234
Ethyl acetate, 109
Ethyl alcohol, 13, 126–127, 139–140. See also
    Ethanol
Etiologies/etiological factors, 180, 183, 189,
    205, 222
    of autism/familial autism, 191, 195–196, 198
    of mental disorders, 218
    neurogenic, 213–214
    prenatal, 180
    of schizophrenia, 220–221
    of shared-environment influences, 217
Eugenics, 61, 237, 296
Eugenics redux, 305
Europe/European, 14, 30, 34, 120, 128, 183, 268
European Union, 275
Evolution, 11, 12, 17, 54, 59, 103. See also
    Darwin's theory of evolution/
    Darwinian evolution
    and brain development, 73, 78, 90, 281
    and epigenetic inheritance, 94
    and ethanol, 129–130
    and fetal programming, 121–122
    heredity and, 202
    and human prenatal development, 59–60
    mechanics of, 61–64
    and natural selection, 75
Evolution Group (mountain range), 59–60
Evolutionary "new brain"/"old brain", 11, 78
Evolutionary biologists, 60, 64
Evolutionary developmental biology, 61, 63–64
Evolutionary psychology, 11, 303
Evolved fetal programming, 121–122
Executive brain function, 137
Executive function deficits, 138
Exogenous chemicals, 118–119
Exophenotypes, 95
External environment, 91, 247
Eyeblink conditioning/eyeblink reflex, 140

Facial defects, 100
Fallopian tubes, 91
Familial autism, 194
Familial clustering, 174
Familial crime, 295–267
Fantasy, 210–211, 280
FAS children, 13, 126, 131–132, 138, 139
    juvenile delinquency predisposition, 138

reduced brain volume in, 139–140
U.S. prevalence rate of, 136–137
Fatty acids/fatty acid chemistry, 99
Feedback mechanisms, 222–223, 283
Female children, 291
Female fetuses, 151, 165, 231
Female homosexuals, 166
Female reproductive system, 151
Female-to-male transsexuals, 144, 152
"Femme" lesbians, 163
Fertilization, 12, 44, 46, 50, 55–57, 91
Fertilized ovum (zygote), 12, 50–52, 55, 56–57, 91, 105–106
Fetal abnormalities (alcohol-related), 132, 135–136
Fetal adrenals, 6
Fetal alcohol disorders. See FAS children
Fetal alcohol exposure, 130, 137–141, 299
Fetal alcohol spectrum disorder (FASD), 13, 58–59, 130–132, 141, 181, 186, 264
    incidence in United States, 136–137
    and language deficits, 299
    and trouble with the law,   138
Fetal alcohol syndrome (FAS), 13, 126, 131–132, 137–140
Fetal androgens, 163
Fetal anomalies/fetal malformations, 20–21, 100, 116, 131
Fetal blood, 6, 221, 278, 283
    alcohol in, 133–134, 139–140
    cortisol level in, 6, 283
Fetal brain, 32, 65–67, 100–101, 133, 166, 175, 177, 200–201, 224, 287
    development of, 73–84, 245, 258
    impact of prenatal factors on, 219–222, 240–241, 243, 277–278, 280, 301
    neurons in, 99
    plasticity of, 68–69
    wiring of, 70–72
Fetal brain damage, 100, 222
Fetal brain development, 67–68, 278
Fetal brain hypoxia, 277
Fetal brain injury, 174
Fetal damage, 21–23, 120–122, 132–137, 275–277, 293
Fetal development, 10, 13–15, 46, 51, 73–79, 81, 82, 86–89, 149, 160, 185, 225
    and development of behavior, 96

disruptions/disruptive effects
    causes of, 108–123, 127, 137
    influences on, 98–103
effect of cultural factors on, 268, 270, 274–279
environmental impact, 25–27, 47, 50, 266–267, 270–273
impact of socioeconomic status, 223, 288–295
maternal behavior and, 280–287
and monozygotic (MZ) twins, 160, 165, 217
season of birth effect on, 218–219
Fetal effects, 5–7
Fetal environment, 20, 223–225
Fetal environmental impact. See Environmental impact
Fetal growth, 92–93, 98–100, 131, 225, 268, 276, 284–285, 290
Fetal growth retardation, 276–277, 285
Fetal hangover. See Alcohol consumption/alcoholism
Fetal hyperthermia, 100
Fetal hypokinesis, 277
Fetal hypothalamus–pituitary–adrenal axis, 223–224
Fetal hypoxia, 100, 198, 219–222
    effect on brain, 100, 101–102, 220
    smoking during pregnancy and, 276–277
Fetal impacts, 21–22, 89, 213, 274, 276–277, 288, 307
Fetal infection, 201, 219–222
Fetal inflammatory responses, 175–176
Fetal malformations. See Fetal anomalies/fetal malformations; Fetal damage
Fetal nervous system, 40, 52–55, 57, 81, 101, 127, 240, 256, 266, 269, 279, 287
    brain-development, 73–74, 78, 89, 205
    mother and fetus, 98
Fetal neurodevelopment, 31–32. See also Neurodevelopment/neurodevelopment disorders
Fetal nutrition, 87–88
Fetal physiology, 88
Fetal polyglobulia, 277
Fetal programming, 6–7, 87, 88–89, 97, 121–122. See also Barker hypothesis; "Prenatal programming"
    and mood disorders, 224, 229
    role of placenta in, 92–93

Fetal sexual development, 145–146, 148–150, 161
Fetal sexuality, 150–151, 152–153
Fetal tachycardia, 277
Fetal teratogens, 126, 217
Fetal testosterone/testosterone, 152
Fetal tobacco syndrome (FTS), 276–277
Fetal toxicity, 105–106
Fetal valproate embryopathy, 194
Fetal vulnerability, 127, 200
Fetal–maternal environment, 107–122
Finger lengths/finger-length ratios, 163, 164, 165
Floods, 6, 7
Flynn, James R., 257
Flynn effect, 255–259, 293
Folate/folate supplements, 279–280
Foreign-born blacks, 268
Fragile X syndrome, 194
France, 257, 300
Fraternal birth order effect, 165–166
Fraternal twins, 194, 243–244, 247, 248, 249
Freudian psychologists, 303
Frontal lobes (of brain), 137, 198
Functional magnetic resonance imaging
    (fMRI), 37

Galen, 149
Galileo, 10
Galton, Francis, 237
Gamma-amino butyric acid (GABA), 215
Gender development, 156–160
Gender-related behavior, 152
Gene and/or environment-dependent
    neuroendocrine alterations of sexual
    brain organization, 166
Gene complexes, 72, 130, 195–196
Gene expression, 12, 20, 49–52
Gene mutations, 82–84
Gene regulation, 12, 50, 51. See also Epigenetics
Gene regulators, 11, 48, 304–305
Gene–environment interactions, 51, 95, 196,
    216, 240
Gene-mongering, 193–196
General intelligence, 236, 237
Generations, 7–9, 54, 60–64, 109, 234–235,
    237, 303
    culture and, 264–266
    differing assessment criteria, 225, 237–240,
        255, 257–260

poverty cycle, 292–293
programming principle in fetuses, 286–287
Genes, 11–12, 23–24, 35, 46–51, 53, 184, 195–196,
    198, 207–213, 217–219, 221–231,
    304–305. See also Gene expression; Gene
    regulators; Genetics; Inherited genes
cerebral cortex and, 89
genotypes and phenotypes, 95–97
and human behavior, 35, 72–73, 198,
    214–216, 220, 304–305, 313n.8
and morality, 301–303
and racial differences, 306–307
sexuality and gender identity, 145–146,
    155–160, 163–167
switching on and off, 62–64, 90–93, 198
and twin studies, 16
Gene-switching cascade, 62–64
Genetic determinism/genetic determinists,
    62–64, 184, 196, 236, 243–244,
    248–250, 254, 306
and criminality, 297–298
and gender, 160
and psychosis, 214–215
Genetic differences/gene-derived variances,
    14–15, 93, 217–218, 239, 256
Genetic disorders, 165
Genetic entities. See Genes
Genetic evolution, 302–303
Genetic factors, 20–21, 26–27, 191, 194
Genetic information, 47–49
Genetic mutation, 174, 183–184, 215
Genetic neuropsychiatry, 214–215
Genetic programming, 145–146
Genetic-determinist psychometricians.
    See Psychometricians/genetic-
    determinist psychometricians
Genetic-determinists, 236, 243, 306–307
Genetics, 11, 20, 46–49, 81, 91, 96, 198, 202. See
    also Behavior genetics/behavioral
    genetics
    familial clustering and, 174
    natural hormonal programming and, 152,
        158, 183
    and schizophrenia, 216
Genital formation, 150–152, 161–162
Genital identity, 144–145, 151
Genital sexuality, 160–162
Genitals, 144–145, 147–148, 151, 154–155, 160–161

Genome identity, 160, 217, 245
Genomes/human genome, 46, 49–52, 56–59,
    62–63, 94–96, 184, 218, 312n.9
    cell differentiation, 90–93
    description of, 47–48
    and identity, 160, 217, 245
    inherited, 218, 244
    and twins, 164–165, 178–179, 184, 248
Genotypes. *See* Genomes/human genome
Georgia, 171, 289
Germ line, 119
Germany, 85
Gestation/early gestation, 44–46, 73, 86, 184,
    235–236, 267–269, 276–278, 288, 305
    cascades of development processes in,
        63–68
    cell proliferation during, 57–59
    consequences of brain damage in, 172,
        174–175, 213, 218–220
    development of gender identity/sexuality
        during, 145–154, 159–160
    developmental periods in, 90–92
    "epigenetic events" in, 52–53, 89
    gene interactions/responses in, 48–49,
        96–97, 104–105, 216–217, 240
    homeostasis principle in, 283–284
    maternal habits during, 126–129, 135–137
    mother and fetus functioning during,
        98–103
    and twins, 194, 245
"g-factor" (general intelligence), 236–237
Gleason, Jackie, 23
Glucocorticoids/glucocorticoid signaling, 101,
    224, 227, 286
Glutamate receptor, 37
Glutamate synapses, 37
Goal-oriented behavior, 137
Gonadal sex determination, 119
Great Britain, 257
Great Lakes, 118
"Ground Zero", 4–5
Group behavior, 54
Group selection. *See* Natural selection
Group sex, 146
Growth retardation, 131, 276, 285

Haeckel, Ernst, 60–61
Hallucinations, 210–212

Handicaps, 27, 49
Hard-wiring, 11, 52, 69, 70, 78, 183, 302,
    304–305. *See also* Brain wiring
Head injury, 186
Hearing impairments. *See* Hearing/hearing
    loss/hearing deficits
Hearing/hearing loss/hearing deficits, 12, 25,
    173, 179–180, 188. *See also* Deafness
Heart disease, 7, 282
Heart rate, 102, 226
Heme synthesis, 39
Hemiplegia, 173
Hemispherectomy, 68–70
Hemostasis (stoppage of bleeding), 154
Hepatotoxins, 127
Hereditarian psychometricians, 35–36, 184, 233.
    See also Psychometricians/genetic-
    determinist psychometricians
Hereditarians, 233, 244, 247, 249–250, 254
Hereditary criminality, 296–297
Heredity, 22–23, 163, 202, 216–218, 239, 296,
    298. *See also* Heritability
    IQ and, 249–250, 254–255, 260, 293
    MZ (identical) twins and, 194–195
Heritability, 35, 216–217, 221–222,
    225–226, 235–237, 238–239, 241–242,
    296, 301, 302. *See also* Heredity
    of behavior and IQ, 35–36, 235, 249–250
    of brain size considerations, 240
    of fetal environment, 254–260
    of genetics/genetic mutations, 95–97, 145,
        174, 215, 240
    of genomes, 164–165, 184, 218
    of identical and fraternal twins, 230–239,
        243–253
    of inherited biochemical information, 50
    Mendelian pattern and, 195–196
    poverty and, 288–289
    prenatal environment and, 241–243
Heterocyclic amines, 278–279
Heterocyclic hydrocarbons compounds, 116
Heterosexual cotwins, 164–165. *See also* Cotwins
Heterosexual males, 163, 166
Heterosexuality, 148–150, 162–167
Hippocampus, 52, 100
Hippocrates, 176
Hispanic children, 266–267
Hispanic teenage girls, 295

Hispanic women (less acculturated), 135
Histological abnormalities, 113
Hoffman, Dustin, 204
    Rain Man (film), 204
Holland, 7, 86
Hollywood, 128
Homeostasis, 230, 283
Hominids, 129
Homosexual lifestyle, 163
Homosexual males/male homosexuals, 166
Homosexuality, 148–150, 160–161, 163–167
Homunculus, 8
Hormonal programming, 152
Hormonal regulation, 150–151
Hormones/hormone secretion, 16, 48, 51–53,
    92, 97–99, 102, 144, 229, 231. *See also*
    Sex hormones; Stress hormones
    brain gender and, 160–163
    endocrine disruptors, 118–119
    and HPA-axis, 285–287
    hypothalamus and, 281–282
    and mood disorders, 224–225
    steroid, 101
House of Commons (Britain), 132
Hudson River, New York, 116
Human behavior patterns. *See* Human
    behavior
Human behavior, 11–12, 19, 22, 24, 50, 72, 130,
    193–196, 254, 298, 301–303
    determinants of, 52–55
    effects of sex hormones on, 152–154
    intelligence and, 18, 41
    measurements of, 80, 235–237
    patterns, 166
    sexual orientation and, 166–167
Human beings, 7, 12, 45–46, 53–55, 80
Human biology/biology, 19, 23–24, 60, 63–64,
    161, 167, 196, 214, 304–305. *See also*
    entries on Biology
    behavior, 298–299
    behavioral genetics, 238–239
    human twins, 245, 249
Human characteristics, 153–154
Human development, 7–12, 63
Human embryonic development. *See*
    Embryonic development/human
    embryonic development
Human evolution. *See* Evolution

Human fetus, 31–32, 48–49, 63, 107, 117, 119,
    120–122, 151–154
Human genome. *See* Genomes/human
    genome
Human morula. *See* Morula
Human ovum, 45–46, 50–52, 105–107
Human qualities, 8–13
Human reproduction, 7, 8, 10, 45
Human traits, 11, 239
Hunger Winter. *See* Dutch Hunger Winter/
    Hunger Winter
Hyperactivity spectrum disorder, 185. *See
    also* Attention deficit hyperactivity
    disorder (ADHD)
Hyperglycemia, 286
Hypertension/postnatal hypertension, 87–88,
    100, 101, 282, 286, 294. *See also* Blood
    pressure
Hypomelanotic skin condition
    (oculocutaneous albinism), 188
Hypothalamic-pituitary-adrenal (HPA) axis,
    281–283, 287
Hypothalamus, 281–282
Hypothyroxinemia, 174
Hypoxia. See Fetal hypoxia

Identical twins, 16, 158, 178–179, 189, 194–196,
    216, 244–254. *See also* Monozygotic
    (MZ) twins
Ideology, 61, 295
Illinois, 170, 171
"Immaculate gestations", 235, 305
Immune system, 47, 81, 101, 175–176, 200, 282,
    312n.9
Immune system protein, 81–82
Immunotoxins, 200–201
Impulse-control disorders, 208
Impulsivity, 185, 277
In utero exposures, 31, 97, 117, 120, 127, 171
In vitro fertilization (IVF), 43–45
India, 30, 34, 257
Individual behavior, 54
Individuals. *See* Phenotypes
Industrial chemicals, 40. *See also* Chemicals/
    chemical elements
"Industrial pollution"/industrial pollutants, 272
Industrialization. *See* Industrialized countries/
    industrialized world

Industrialized countries/industrialized
world, 3, 32, 39, 97, 113–114, 116, 242,
289–290, 300
Infancy, 6, 53, 81, 113, 172, 225, 285
gender identity during, 155, 157
short-term memory deficits in, 117–118
Infant mortality, 120, 245, 266–269, 270–271
and geographical differences, 272–273
maternal health care during pregnancy and,
274–280
Infant temperament, 223, 226–227, 285–287
Infants, 5, 30, 45, 76, 132, 138–139, 158,
302–303
BLL and, 31–32
and congenital anomalies, 178, 179
low birth weight in, 225, 267, 268, 270, 273,
274–275, 288
temperaments of, 223, 226
Infections, 15, 53, 100–101, 120, 175, 178, 180,
186, 293
and autism syndrome, 200–201
and fetal impacts, 219–223, 274, 294
respiratory tract, 270–271
and season of birth, 218
Inflammation/inflammatory agents/
inflammatory responses, 101,
175–176, 222
Inflammatory intestinal response.
See Inflammation/inflammatory
agents/inflammatory responses
Influenza infections. *See* Infections
Inherited genes, 54–55, 196, 238, 240, 244, 247,
249–250, 254
Inherited traits, 235–237, 296–298, 301
Insulin-like-growth factor (IGF), 97
Intellectual impairment/intellectual behavior,
40, 117–118. *See* IQ/IQ tests
Intelligence quotient (IQ). *See* IQ/IQ tests
Intelligence. *See* IQ/IQ tests
Intra-uterine growth. *See* Intrauterine growth
restriction (IUGR)
Intrauterine growth restriction (IUGR), 5–7,
99, 270–271, 277
Iodine deficiency. *See* Iodine nutrition/iodine
deficiency
Iodine nutrition/iodine deficiency, 53,
175, 200
Ionizing radiation/radiation, 39, 106, 284

IQ/IQ tests, 14, 17, 25, 29, 127, 181, 216–217,
233, 251, 271. *See also* Heredity;
Heritability
brain size and, 240–241
evolution of techniques for, 234–240
limitations of, 256–260
prenatal environment and, 241–243
Isomorphism, 147
Isooctane, 109
Israel, 257

Jamaica, 30, 272
Japan/Japanese, 108, 110, 114, 203, 257, 270
Jena, Germany, 60
Johns Hopkins Medical Center, Baltimore,
155, 156
Journal of the American Medical Association (JAMA),
15
Jukes family, the 296
"Junk DNA", 47, 48, 49, 52

Kael, Pauline, 187
Kenya, 257
Kidneys, 82

Labels, 153–154, 161–162, 177, 202, 207–209,
225, 232, 304
constellation of symptoms, 190–191
IQ test performance, 259–260
psychiatry, 211–213, 214–215
Lancet, The, 43, 131, 132
Language constructs, 147, 148–149
Language deficits, 299
Language function (of brain), 37, 68
Lazio, Italy, 264
Lead exposure/prenatal lead exposure, 15–16,
26, 31–32, 37
and ADHD, 37–38, 186
cause cognitive deficits, 28–29
and criminal behavior, 299–300
fatality, 26–27
low-level, 38, 185
and schizophrenia, 221–222
Lead poisoning, 28–36, 37–39, 186, 300
Lead toxicity. *See* Environmental lead
pollution; Lead poisoning
Leaded gasoline, 32–33, 265
Learning abilities, 37, 183, 190

Learning disabilities, 138–141, 169, 171, 180–185.
    See also Dyslexia
Learning problems/behaviors, 26–27, 81. *See
    also* Dyslexia
Learning/learning experiences, 26, 37, 66–68,
    81, 146, 155, 180
    behavior patterns, 82, 183
    memory and, 70–72, 78
    postnatal environment, 53, 163
    transgenerational passage of, 54–55
Lehrman, Daniel S., 23
Life span, 45–46, 53
Lifetime/lifetime prevalence, 16, 170, 177, 208,
    211–213, 228–229
Ligands (specific binding and interacting
    molecules), 99
Limb defects, 100
Linearizations, 239–240
Lipids (cell-membrane fat molecules), 56, 99
LIS1 (gene), 82
Lissencephaly, 82–83
Live birth, 46, 136–137, 169, 170–171, 267,
    268, 276. *See also* Low-birth-weight
    children
Live newborns/live-born children. *See*
    Newborns
Live-born children. *See* Newborns
Liver cells, 45, 49
Local chemistry, 77, 79, 177
Local culture, 22, 265–266
Local environments, 48–49, 52–57, 62,
    145–146, 272
    brain development, 67–68, 77
    effects on gene expression, 75–76
    uterine environment, 78, 93
Lombroso, Cesare, 296
Long-term consequences/long-term effects
    (for health), 15, 97, 102, 117
Long-term memory, 117
Los Angeles, 29, 135
Low-birth-weight children, 13, 25–27, 100, 112,
    120, 266–269, 271–274, 276, 288. *See
    also* Birth weight
    Barker hypothesis, 87–88
    fetal environment and, 25–26, 223–232,
        270–271
    maternal health care during pregnancy and,
        274–280

Lower socioeconomic classes, 178, 256, 293
Low-level lead exposure. *See* Lead exposure/
    prenatal lead exposure
Loy, Myrna, 128
Lung growth, 120, 270
Lyme disease, 201

Macromolecules, 77
Macrophages (immune system scavenger
    cells), 175
Madness, 209–210
Magnetic resonance imaging (MRI), 37, 102,
    139–140, 177
Major depression, 208, 218, 223, 228–229, 231.
    See also Major depressive disorder
Major depressive disorder, 208, 228–229, 231.
    See also Major depression
Male children, 291
Male fetuses, 148–149, 151–152, 165–167
Male infertility, 119
Male-linked antigens, 166
Male-to-female transsexuals, 144, 162
Malnutrition, 53, 85, 86, 230
Manhattan, 4, 36, 125, 128, 290
Manic-depressive illness, 212
Man-made effects, 3, 21, 22, 171–172, 284, 307
March, William, 297
    The Bad Seed (novel), 297
Marijuana, 39
Marine worms (Bonellia), 146
Massachusetts, 30
Maternal alcoholism. *See* Alcohol
    consumption/alcoholism; Pregnant
    mothers; Pregnant women
Maternal anemia/anemia, 231, 237, 294
Maternal behavior (prenatal and postnatal),
    101–103, 269, 274–280
Maternal BLL. *See* Blood lead concentrations;
    Blood lead level (BLL) standards
Maternal blood alcohol. *See* Pregnant mothers
Maternal dichloro-1. diphenyl-trichloroethane
    (DDT) exposure, 185
Maternal environment, 31–32, 51–52, 92–93,
    258, 285, 301
    cultural factors and, 285–287
    environmental toxins, 107–122
Maternal hypertension, 100
"Maternal immunity theory", 166

Maternal infection, 100–101, 178, 180, 220, 293
Maternal instinct, 304
Maternal iodine deficiency, 175, 200
Maternal lead values. Blood lead concentrations; Blood lead level (BLL) standards
Maternal metabolic dysfunctions, 175
Maternal nutrients, 57, 63, 247
Maternal physiology, 98, 280–281
Maternal psychological stress, 5–7, 220–222
Maternal smoking, 185, 231, 277. *See also* Tobacco smoke/effects
and bulimia nervosa in offspring, 231
and criminal tendencies of offspring, 300–301
Maternal stress during pregnancy, 15, 221–225, 270, 284–286
Maternal tobacco smoking. *See* Maternal smoking; Tobacco smoke/effects
Maternal under-nutrition, 121–122
Media ballyhoo/media carnival/media myth, 22, 23, 46–47, 130, 149, 202, 247, 255, 264–265, 278, 295, 302, 303
autism, 193–194, 196, 204
ethnic groups and intelligence, 235
psychometrics, 137
Medical Hypotheses (journal), 296
Medical journals, 5, 43, 44, 131, 155
Medical negligence/malpractice, 173
Medical psychiatry, 211–213
Megalomania, 212
Mendelian genetics, 195–196, 298
Mendelian laws of heredity, 216
Mendelian traits, 195–196, 216
Meningitis, 186
Menstruation, 230
Mental deficits, 126
Mental disorders, 17, 199, 207–224, 230–232, 292
Mental illness, 86–89, 207, 209–222, 229
ambiguities in diagnosis of, 209
higher order dysfunctions of, 230–232
poverty and, 291–292
prevalence, 208
temperament and moods, 223–228
Mental retardation. *See* Mentally retarded/mentally retarded children

Mentally retarded children. *See* Mentally retarded/mentally retarded children
Mentally retarded/mentally retarded children, 29–30, 36, 140–141, 241
Mercury/mercury compounds/mercury pollution, 10, 22, 53, 108, 109–112, 114–115, 199, 242. *See also* Methylmercury/methylmercury poisoning
Metabolic syndrome, 121
Metabolism/metabolites/metabolic rates, 57, 87–88, 97, 98, 121, 170, 195, 276–277, 282
Methoxychlor, 119
Methylmercury/methylmercury poisoning, 39–40, 109–115, 171–172, 174, 176, 203–204. *See also* Mercury/mercury compounds/mercury pollution
Mexican-American women, 268
Mexico City, 28, 31
Michigan, 170
Microcephaly, 140
Microdeletion (of genes), 82–84
Microorganisms, 101
Microtubules, 77
Middle Ages, 148
Middle classes, 118, 251, 253–254
Middle East, 34
Migrations, 48, 56–57, 83
of nerve cells, 76, 78, 80
neuronal, 77, 78, 99, 139
Minamata City, Japan, 108, 109, 110, 112, 113, 203
"Minamata disease", 110, 112–113
Minor depressive (dysthymic) disorder, 228
Minorca Island, Spain, 97
Mirror neurons, 197–198
Modern society, 32, 40
Molecular biology, 11, 74, 95, 280–281, 298
Molecular neuropsychiatry, 209
Molecules, 8, 11, 48, 56, 99, 105–107
Monochorionic MZ twins, 246, 248
Mononucleosis, 282
Monozygotic (MZ) twins, 184, 194. *See also* Fraternal twins; Identical twins; Twin research/studies
autism concordance in, 194
cognitive performance of, 250

Monozygotic (MZ) twins (*cont.*)
   environmental involvement, 244–246
   heritability of IQ, 249, 251–254
   maternal nutrient environment, 247–248
   role of heredity, 195–196
   schizophrenia in, 216–218
Mood disorders, 225–227
Morality instinct/moral instincts/moral
   sense, 301–305
Morality. *See* Morality instinct/moral
   instincts/moral sense
Morphological disruptions, 59–60
Morula, 44, 91, 105–106. *See also* Blastocysts
Mother and fetus function, 98–103, 284
Mothers, 14, 17, 112–113, 276, 303. *See also*
         Pregnant mothers; Pregnant women
   adoption studies, 253–254
   autistic children, 192–193, 199, 200, 204–205
   DNA adducts of, 270
   folate dietary insufficiency of, 279–280
Motor control, 40, 172
Motor cortex, 172, 173, 176
Motor disorder, 173
Mt. Sinai School of Medicine, 4, 5, 6
Muscles, 172, 176
Mutagens, 119–120, 278–279
Myelination of nerve fibers, 93, 172

Narcotics, 102
"Narrow" autism/severe autism, 187–188, 189
National average, the/national averages, 27–32
National Book Award, 297
National prevalence, 170–172
Natural selection, 11–12, 63, 75, 129. *See also*
         Darwinian natural selection
Nature, 18, 49, 147
Nature–nurture debate, 12, 18, 54
Nazis, 61, 85, 192
Negative temperaments/negative
   emotionality, 226–227
Neonatal defects 170–172
Nephrotoxins, 127
Nerve cell connections, 37, 80–81
Nerve cell maturation, 73, 83
Nerve cell migration. *See* Cell migration/nerve
   cell migration
Nerve cells, 12, 58–59, 99, 107, 126, 139, 176,
   197, 198, 214, 241

brain lobes and, 71–78
   interaction between, 82
   and local biochemical events, 80–83
Nerve fibers, 93, 139, 172, 281
Nerve impulses, 126, 197–198
Nervous system, 7, 13, 17, 32, 40, 53–55, 57,
   73–74, 89, 98–99, 220, 266, 269,
   279, 287
   and autism, 205–206
   changes in, 240–241
   hard-wiring, 52, 78, 81
   prenatal and postnatal damage to, 14–16,
   101, 256
Netherlands, the, 85, 257, 259
Neumann, Heinrich, 209
   "loosening of the togetherness", 209
Neural axons, 80, 100
Neural circuits/neural circuitry, 40, 77
Neural connections, 53, 93, 133
Neural development/fetal neural
   development, 17, 31–32, 81–82
Neurobehavioral changes/neurological
   impairments, 39, 118, 299
Neurobiologists, 23
Neurocognitive deficits, 120, 182–183, 271
Neurodegenerative diseases, 81–82, 87–88
Neurodevelopment/neurodevelopment
   disorders, 16, 31, 95–98, 131–132, 215,
   220, 229
   environmental agents, 40–41
   metals, 115
   prenatal lead exposure, 31–32
   schizophrenia, 215, 220
   subclinical neurotoxicity, 115
Neuroendocrine abnormalities, 221
Neuroendocrine system, 98, 221, 224
Neurogenesis, 69–70
Neurologic symptoms/neurological
   symptoms, 111, 113
Neurological ability. *See* Neurological
   development/ability
Neurological damage/deficits, 38–39, 81,
   172–173
Neurological development/ability, 54,
   188, 284
Neurological disorders/difficulties, 137, 172, 189
Neurological seizures. *See* Seizures/
   neurological seizures

Neurological substrates, 227
Neurological wiring, 81. *See also* Brain wiring;
    Hard-wiring
Neuromotor system, 180–181
Neuron cytoplasm, 78
Neuron density, 79
Neuron membranes, 126
Neuronal circuitry, 21, 99
Neuronal connections, 68
Neuronal death, 100
Neuropsychiatry/neuropsychiatric disorders,
    173, 214
Neuropsychological deficits, 139, 140
Neuropsychological developmental disabilities,
    169–171, 193–205
    involving prenatal environmental impact,
        172–187, 191–192
    prevalence of, 188–190
Neuroscience, 71, 73, 266
    contribution of psychometrics to, 236–237
    fetal brain development, 74–83
Neuroscientists, 54, 68, 84
Neurosurgeons, 68, 69
Neurotoxic action. *See* Neurotoxic effects/
    action
Neurotoxic effects/action, 35, 40–41, 115,
    242. *See also* Neurotoxins; Prenatal
    toxicology; Toxicity/toxic effects
Neurotoxins, 25, 37, 105–106, 107–108, 127, 129,
    288, 289–290, 294
    effect on IQ, 242–243
    ethyl alcohol/alcohol, 126–129
    food chain hazard, 112–114
    lead, 107–108, 221–222, 242–243
Neurotransmitters, 82, 95, 102, 198, 227, 229
New Orleans, 36
New realities, 13–16
New York City, 5, 290
New York Times, 297
New Yorker, 187
New Zealand, 257, 300
Newborns, 3, 20, 45, 57, 130–131, 178–179, 225,
    300–304
    damage sustained by, 120–122, 173–174
    mental retardation in, 128–129
    neurobehavioral changes in, 118
Nicotine/nicotine dependence, 101–102, 136,
    292, 301

"9/11", 4, 6, 213, 281
NMDA (*N*-methyl-d-aspartate) receptor,
    37, 38
nongene DNA/"junk DNA", 47
Nongenetic differences, 14
Non-Hispanic-American (highly acculturated)
    women, 277
Non-Hispanic black women, 135, 200, 295
Non-Hispanic white women, 135, 271,
    272, 295
Non-Mendelian complex traits, 72
Noninfectious toxins, 175–176
Nonpsychotic mental illness, 223–231
North America, 34
Northern Ireland, 292
Norway, 16, 257
Notochord, 61
Nucleohistone/histones. *See* Chromatin/
    chromatin proteins
Nutrition. *See* various entries for Nutrition
Nutritional deficiencies/deprivation, 15, 178,
    219, 222

Obesity, 121
Obstetrical delivery, 173–174
Obstetricians, 4, 44, 276
Occupational exposure, 40
Oculocutaneous albinism (hypomelanotic skin
    condition), 188
Offspring, 38–39, 96, 102, 115, 152, 231, 277, 286,
    300–301
    advanced maternal and paternal ages
        and, 199
    effects of poverty on, 292–293
    maternal alcohol consumption and,
        132–137, 140, 243
    mental illness among, 86
    prenatal exposure impact on, 185–186,
        200–201, 284, 290
    pseudohermaphroditic, 118
    schizophrenia in, 219–221, 285
Offspring behavior, 96, 102, 285. *See also*
    Autism; Autistic children
"Old-brain" behaviors/function, 78, 79
Ontogenesis. *See* Development (ontogenesis)
Opposite-sex twin pair, 231
Ordinary behavior, 208–209
Organisms, 12, 20, 51, 298

Organochlorine pesticides. *See* Pesticides/ organochlorine pesticides
Organs, 20–21, 56, 88–89, 93–94, 120, 144

Parental procreation habits, 218
Parents, 9–11, 18–19, 27, 150, 204–205, 243, 288
  autism spectrum disorder (ASD), 199–201
  autistic behavior, 187–188
  Bettelheim's dogma, 192–193
  dysfunctional temperaments, 226–227
  human sexuality continuum, 153–160
Parkinson's disease, 88
Particulate organic matter, 119
Pathogens, 174, 274
Pavlovian reflex, 140
Pearson correlation coefficient, 252–253
Pearson, Karl, 252
Pediatric psychiatry, 185
Pediatricians, 29, 36, 114, 183
Pediatrics, 4, 6–7
Penile ablation, 154–158. *See also* "Ablatio penis"
"Penis envy", 149–150
Penis, 149, 151, 154, 155–159
Peripheral muscles, 172
Peripheral nervous system, 127
Periventricular leukomalacia (brain lesions), 172–173
Personal responsibility, 22–23
Personality traits, 72, 226–227
Pervasive development disorder (PDD), 189
Pervasive developmental disorder not otherwise specified (PDD-NOS), 189
Pesticides/organochlorine pesticides, 39, 200, 290
pH levels, 33
Phalloplasty, 159
Pharmaceuticals, 102–103
Pharyngeal pouches, 61
Phenomenology, 194
Phenotype behavior, 96, 186
Phenotypes, 46, 50, 62–64, 94–96, 100, 186, 195–196, 239, 298
  behavior of, 186
  cellular toxicology of, 186
"Phenotypic variation", 93
Philadelphia, 36
Phimosis, 154

Physical environment, 53
Physiology/physiological, 6–7, 23, 63, 87, 95, 102, 118–119, 145, 228, 245, 268, 280–284, 303–304
  alterations in organ systems of, 88–89
  differences between MZ twins, 84
  fetal brain, 201, 220
  homeostasis, 230
  impact, 174
  neuroendocrine systems, 224, 287
  programming, 286
  stress, 92, 224, 286
  systems, 7, 283, 284
  temperament, 226–227
Placenta, 6, 31, 92, 97, 102, 231, 277, 285
  detachment of, 100
  and development of twins, 16, 93, 158, 246
  in interaction between mother and fetus, 51–52, 98–99, 112–113, 120, 133–134
  in maternal nutrition, 276, 279
  progesterone supply by, 283
  and X-chromosome inactivation, 16
Placental functional deficit, 99
Placental genes, 92–93
Placental insufficiency, 98–99
Placental metabolism, 97
Plasma glucose, 224, 286
Plasma levels, 31
Plasma lipid concentrations, 224
Plasticity. *See* Brain plasticity
Polar bears, 118
Polychlorinated biphenyls (PCBs), 22–23, 39, 118–119, 171–172, 242
  cognitive/learning abilities resulting from, 183
  immunotoxic effects of, 200–201
  impossibility of eliminating, 115–117
Polycyclic aromatic hydrocarbons, 120
Polypeptide protein hormone, 97
Polyphenolics, 90
Poor children, 14, 256, 291–292
Poor families, 14, 255, 270
Postnatal behaviors, 89, 141, 183, 282, 290
Postnatal damage, 14, 256
Postnatal environment, 14, 18, 53, 93, 157, 217, 269, 290
  behavior traits, 160, 306
  and identical twin pairs, 244, 248–249, 252–254

and phenotypic differences, 93–94
and sexual orientation, 163–167
Postnatal learning processes, 163
Postnatal psychosexual development/
psychosexual development, 152–154,
165–167
Postnatal seizures (epileptic), 178
Postnatal temperament, 88–90, 226–227
Postnatal theory, 191–193
Poverty, 22, 118, 133, 180, 181, 302–307
adolescent pregnancy, 294
effects of living conditions/environment,
289–293
health care implications of, 288–295
inherited disease/criminal behavior, 288,
295–301
psychological stress of, 281–283
Powell, William, 128
Pregnancy. See Pregnant mothers; Pregnant
women
Pregnant mothers, 13, 117–118, 127, 132, 149–150.
See also Pregnant women
Pregnant women, 4–5, 13, 22, 219–220, 274,
280. See also Pregnant mothers
alcohol consumption by, 22,
128–130, 132, 133–134, 136, 141, 243,
264–265
coffee-drinking by, 278, 293
drug use by, 102
environmental lead pollution/air pollution,
31–32, 36, 38, 120
exposure to pesticides of, 200
folate supplements, 279
Lyme disease and, 201
tobacco use by, 292
Preimplantation embryos, 90, 91
Premarin, 157
Premature delivery/preterm
delivery, 277, 294
"Premature" infants. See Low-birth-weight
children
Prenatal androgens/prenatal androgen
exposure, 163
Prenatal brain, 82, 140, 175–176, 215
Prenatal brain development. See Brain
development/fetal brain
development
Prenatal damage/disruptions, 13, 14, 18

Prenatal development/human prenatal
development, 12, 17, 20–21, 46, 54–55,
73–75, 101, 112–113, 121, 145
and developmental aberrations, 79–80
evolution and, 59–64
and sex hormones, 150–154
vulnerability of, 55–58
Prenatal environment, 7, 12, 13–18, 61, 62, 80,
88, 250, 254–255
and behavior and intelligence, 52–53, 160
causes of pollution and, 21–23, 39, 41
and embryo implantation, 92–93
global, 62–64
impact on neuropsychological
developmental disabilities of, 171–172
and IQ measurement, 35–36, 233, 241–249,
306
lethal and nonlethal effects of, 92–93
and MZ twins, 194–196, 217–218,
250–254
and postnatal effects, 61, 88, 96, 117,
183–184, 306
prenatal damage and, 13–18
and psychological sexuality, 145–146
and schizophrenia spectrum disorders
(SSDs), 219–222
and sexual orientation, 163–167
Prenatal environmental impacts/effects,
88–89, 92
Prenatal exposure, 21–22, 181, 220, 221–222,
224, 284–287, 291
to alcohol, 138–140
and effect on cognitive and learning
abilities, 183
effect on IQ, 242–243
to lead, 21
to PCBs, 117, 200
to pesticides, 290
and postnatal ADHD, 185–187
to released androgens, 163
Prenatal growth, 7, 97
Prenatal impacts, 18–21, 70, 177–179, 184, 215,
231–232, 291–293, 300–301
on adult pathology, 221–223
on autistic behavior, 191–193, 198
on cognitive performance, 241–243
environmental toxins and, 171–172
and plasticity in the fetal brain, 70

Prenatal lead exposure. *See* Lead exposure/
prenatal lead exposure
"Prenatal programming", 6, 89, 284. *See also*
Fetal programming
Prenatal toxicology, 103–107
Prenatal vulnerabilities, 46–47, 57–59
Preschool children, 300
Prescription drugs, 186
Preterm birth, 99, 225, 267, 270, 277
Preverbal infants, 302
Primates, 283
Private demons, 210
Prodromal signs (of schizophrenia), 211
Progenitor cells, 91, 127
Progesterone, 283
Prohibition (of alcohol, 1920 to 1933), 128
Proinflammatory cytokines, 100–101, 220
Protandrous hermaphrodites, 146
Proteins, 48–50, 56, 81, 91, 100–101, 175, 283
"Proxy markers", 163
Pseudohermaphroditism, 118–120
Psyche, 4, 187, 295
Psychiatric diagnoses, 190–191, 208–209
Psychiatric disorders, 101, 199–201, 215, 222,
232, 270
poverty and, 289, 291–292
socioeconomic status and, 222–223
Psychiatric illnesses, 95
Psychoactive drugs, 103
Psychoanalysis/psychoanalytic professionals,
192–193
Psychogenic mechanisms, 291
Psychological development, 192
Psychological distress, 229
Psychological dysfunction, 224
Psychological identity, 144
Psychological impairments, 39, 299–301
Psychological stress/maternal psychological
stress, 5, 220–221, 224, 280–281, 291
Psychological tests/testing, 137
Psychometric methods. *See* Psychometrics/
psychometric studies
Psychometricians/genetic-determinist
psychometricians, 235–236
Psychometrics/psychometric studies, 35–36,
41, 183–184, 233–240, 245
Psychopathology, 72
Psychosis, 17, 207–217, 220–222

nonpsychotic mental illness and, 223–232
season of birth and, 218–219
Psychosocial identity, 148
Psychosocial stress/stressors, 198, 283–284,
286–287
Psychosomatic research, 280–281
Ptosis, 131
Public awareness, 187, 204
Public health, 127
alertness to correlations, 203–204
changeability of data, 19–23, 133
Public policy, 54, 239, 258–259, 266, 305,
323n.10
Purkinje cells, 111
Pychosocial gender, 150

Quadriplegia, 173

Racial disparity, 267–269
"Racial" groups, 35, 190
Racism, 61, 237
Raven's IQ score (Raven's Standard
Progressive Matrices), 257
Reading disabilities. *See* Dyslexia; Learning
disabilities
Reality, 8, 210–211, 214, 239
Regulating genes, 47, 62
Regulatory proteins, 100–101, 283–284
Reprogramming, 51–52, 90–91, 119
Researchers/research studies, 5, 7, 13, 16, 62, 87,
89, 104–105, 117, 120, 129, 202, 215–216,
225, 245, 254, 280–281, 306
birth weight categorization, 268
concurrent use of alcohol and tobacco,
136–137
development of autism, 198
discovery of microdeletion, 82–83
exposure to common organic
solvents, 40
IQ, 14, 37, 233–234
lead pollution, 29, 31–32, 38
prenatal exposure to PCBs, 117–118
transgenerational consequences of
famine, 86
Respiratory tract infections, 120, 270
Rett syndrome, 189, 198
Ribonucleic acid (RNA) molecules, 47–49
Ribonucleic acid (RNA), 11, 47–49

Risk factors, 92, 224–225, 231–232, 310n.15. *See also* Lead exposure/prenatal lead exposure
  caffeine, 278
  during pregnancy, 274, 294–295
  lead exposure, 15
  maternal habits during pregnancy, 185, 221–222
RNA product, 48
Rochester, 28
Rome, 22, 264, 265
Rubella (congenital German measles), 194

Scavenger cells (macrophages), 175
Schizoaffective disorder, 212–213, 218
Schizophrenia, 6, 15–16, 21, 86, 199, 208, 209, 210, 211
  fetal environment and, 211–212, 223–224
  possible causes of, 213–222, 285
  prenatal exposure to lead, 21–22, 38–39
  season of birth correlation, 97, 230–231
  violent paranoid, 207
Schizophrenia spectrum disorders (SSDs), 212, 219–220, 221
Schizotypal personality disorder, 212
School-age children, 115, 170, 182
Science (journal), 13, 15
Season-of-birth effect, 177–178, 219, 230–231
2D–4D ratio (second to fourth finger-length ratio), 164–165
Second World War, 11, 85
Second-hand smoke, 102
Sedgwick, Adam, 10
Seizures/neurological seizures, 68, 82, 112, 176–179, 189
Sensations, 176, 190, 230
Sensitivities, 40, 71, 153
Sensory impairments, 180
Sensory responses, 40–41
September 11, 2001. *See* "9/11"
Sex chromosomes, 157, 161
Sex hormones, 102–103, 150–154, 162
Sex reassignment surgery, 143–144, 154–160
"Sex-dimorphic physical characteristic"/sex-dimorphic characteristic, 164–165
Sex-reversal, 118
Sexual behavior, 153–154, 162–166
Sexual differentiation, 161–162

Sexual dimorphism, 119, 162–163, 164
Sexual intercourse, 157, 274
Sexual odyssey, 145–146
Sexual orientation, 148, 162–167
Sexual variations, 166–167
Sexuality, 78, 282
  in animal world, 146–149
  complex interactions in the fetus and, 150–159
  determination of gender identity/sexual orientation, 160–167
Sexually transmitted diseases, 294
Shared-environment etiology, 217
Short-term memory, 117, 285
Siblings, 27, 195, 243, 249. *See also* entries relating to Twins
"Silent pandemic", 40
Singletons, 245
Skin cells, 45
"Small for gestational age" (SGA), 99, 267–268, 276
Social behavior/antisocial behavior, 11–12, 18, 291
Social consequences, 21–22, 291–292
Social drinking (nonbinge), 133, 138
Social environment, 53, 280, 282, 290
Social interactions, 26–27, 188, 190, 302–303
Social learning, 155
Social maladaptation/maladjustment, 299–300
Society, 3, 20, 22, 32, 148, 250, 292, 304
  and fetal damage, 293
  human sexuality in, 153–154
  poverty in, 288
  "silent pandemic" in, 40
Sociobiology/sociobiologists, 11–12, 302
Socioeconomic circumstances, 290–293
Socioeconomic class/status, 18, 222–223, 250, 255, 270, 293
Sonia Schankman Orthogenic School (for emotionally disturbed children and adolescents), 192
South Africa, 30
South Bronx, 290
Southern California, 135, 295
Spain, 97, 257
"Spandrels" (evolutionary free-riders), 130
Spanish-speaking Hispanic women, 295

Spastic cerebral palsy, 173–176. *See also* Cerebral palsy
Spearman, Charles, 236
"Spearman g-factor", 236–237
Specific proteins, 48, 49–50
Speech and language functions/development, 68, 181
Speech deficits/defects, 169, 171, 180–185
Sperm cells, 8, 10–11, 44–45, 55
Spermatogenic capacity, 119
Spina bifida, 100, 136
Stanford–Binet IQ tests, 234, 251, 257. *See also* IQ/IQ tests
"Steerability", 72
Stem cells, 12
Steptoe, Patrick C., 43, 44, 45
Steroid hormones, 101, 283
Stillbirth/stillbirth rate, 276, 285
Stomach, 279
Stress hormones, 6, 224, 225, 281, 282, 286. *See also* Cortisol/cortisol levels
Stress research, 281
Structural defects/anomalies. *See* Fetal anomalies/fetal malformations
Subclinical behavior, 84
Subclinical toxicity, 171–172
Substance abuse, 97
Substance disorders, 208
Substrates:
 biological, 109, 269, 290–291, 298–299
 neurological, 227
 simple genetic, 50
Suicide/suicidal thoughts, 52, 156, 193, 204, 228, 297
Sun, 11
Superfund Hazardous Waste Site, 116
Suprarenal glands, 281
Surgeon General (United States), 134
Survival/survival rates, 45–46, 80–81, 93, 97, 121–122, 130, 153–154, 267, 281
 psychological stress in poverty, 281
 trade-off in, 93
Sweden, 257, 300
Switzerland, 257
Symptom clusters, 209
Symptoms, 36, 109, 195, 208–209, 228, 285
 of ADHD, 186
 of anorexia nervosa, 230

 of autistic/ASD, 27, 190–191, 201
 of *Cytomegalovirus* infections, 178
 of damage to motor cortex, 173–176
 of influenza infection, 220
 of lead poisoning, 36, 38–39
 of psychosis/mood disorders, 212–213, 228
Synapses, 37, 68, 70–71, 80–82
Synaptic plasticity. *See* Brain plasticity
Synaptogenesis, 133
Syndromes, 189, 194, 198
 autism, 200–201
 clinical, 172
  cerebral palsy, 172–173
 Down syndrome, 82, 136, 195
 fetal alcohol syndrome (FAS), 13, 126, 131, 132
 fetal tobacco syndrome (FTS), 277
 fragile X, 246
 metabolic, 121
 MZ transfusion, 246
 schizophrenia, 215–216
Synthesis, 39, 49, 57, 82, 109, 165
Synthetic chemicals, 15, 102
Synthetic retinoids, 102
Szasz, Thomas, 211

Taiwan, 117
Teenage girls, 294–295
Teenage pregnancy, 277, 294
Temperament/postnatal temperament, 88, 96–97, 225–227
Teratogens, 127, 217
Testes, 151, 155
Theory of "neural Darwinism", 80–84
Theory of evolution, 10, 11. *See also* Darwin's theory of evolution/Darwinian evolution
Thimerosal, 200
"Thin Man" detective films, 128
"Threshold" concentration, 13, 28–30, 105, 133, 242
Thymus gland, 200
Thyroid drugs, 103
Time of birth, 7, 86
Tissues, 12, 20, 34, 52, 87–89, 116
 brain and nervous system, 40–41
 cell differentiation and, 90–91, 93

development, 56–57, 63, 74, 150, 283
  effects of alcohol on, 127, 134
Tobacco smoke/effects, 21–22, 52–53, 100,
    265–266, 271, 276–277, 300–301. *See
    also* Cigarettes
"Tomboyism", 165
Toxic chemicals. *See* Chemicals/chemical
    elements
Toxicity/toxic effects, 13–16, 25, 27, 175, 183,
    290. *See also* Neurotoxic effects/action;
    Neurotoxins; Prenatal toxicology
  of chemicals, 87–88, 110–118
  of ethanol/alcohol, 103, 133–137
  of lead, 27–39
  subclinical, 171–172
Toxin concentrations, 103–104
Toxin molecules, 67–68, 103–104
"Toy scare", 33
Transcription/transcription regulators, 47–50,
    95, 99, 298
Transgenerational cognitive dysfunctions and
    deficits, 306
Transgenerational diseases/disease state, 119
Transgenerational non-DNA inheritance of
    temperament, 96
Transgenerational nongenetic disorders, 7,
    222, 277
Transgenerational nongenetic inheritance of
    psychiatric disorder, 221–222
Transgenerational nongenetic stress disorder, 7
Transgenerational nongenetic tobacco impact,
    277
Transgenerational passage of learning, 54
Transgenerational transmission of phenotypic
    variation, 94
Transsexuality/transsexualism, 17, 144–150,
    159–160
  brain studies of, 153–154
  development of gender identity, 161–162
  discordance in monozygotic twins,
    164–167, 184, 216–218, 245–254
Trauma, 6, 26–27, 177, 138
"Trophic factors", 82
Tumors, 279
Turkheimer study, 255–256
Twin pairs, 14, 231, 244, 249, 250–252
Twin research/studies, 16, 217, 183–184,
    216–218, 237, 244–245

U.S. Centers for Disease Control (CDC),
    28–29, 190
  autism prevalence rate, 190
  IQ deficit, 242
  standards for water purification, 272–273
U.S. Environmental Protection Agency
    (EPA), 29, 36, 114, 299
U.S. population, 135, 208
Ultrasound measurements, 99
Umbilical cord/umbilical cord blood, 31–32,
    100, 114–115, 118, 290. *See also* entries
    for Placenta
Unbalanced nutrition, 121
Unborn, the, 19, 22, 112, 143–144, 147, 153–167,
    264, 288, 307
  beginning of sexual odyssey, 145–146,
    148–150
  formation of genitals, 151–152
Unconditioned stimulus, 140
Undeveloped countries, 45
Unitary entities/elements, 47–49, 298
United States, 25, 26, 30, 32, 116, 128, 144, 169,
    181, 191, 200, 203, 242, 245, 257, 271,
    275–277, 279, 300
  alcohol use, 128, 134–135, 136–137, 183, 264
  antibodies to *Cytomegalovirus*, 178
  autism prevalence in, 190, 196
  commercial use of PCBs, 116
  criminality, 300
  infant mortality rate in, 267–268
  irregular surveys on disease prevalence,
    179–180
  schizophrenia in, 212, 216
  standard for water purification, 272–273
University of Chicago, 192
University of Vienna, 192
Upper classes, 14–16, 133, 178, 256, 288, 292
Urban children, 30–32
Uterine fibroids, 118

Vaccine-mercury linkage (in postnatal
    autism), 199–200
Vaccines, 102, 199, 200
Vaginoplasty, 157
Van Leeuwenhoek, Anton, 8
Verbal communication, 26–27, 187–188, 190
Vertebrates, 162–163
Vinclozalin (antiandrogenic compound), 119

Vinyl acetylene, 109
Violence, 210, 271, 288, 299–300
Violent paranoid schizophrenia. *See*
    Schizophrenia
Virus infections, 178, 180, 186, 220
Visual impairment/vision impairment/vision
    loss, 20, 111, 171, 173, 179–170, 180,
    188. *See also* Blindness; Childhood
    blindness
Voluntary muscles, 173, 176

Watson, James D., 11
Weaver, Sigourney, 204
   Snow Cake (film), 204
Wechsler Adult Intelligence Scale (WAIS),
    257
Wetterhahn, Karen E., 110, 111, 112

Whiskey, 128, 193
White matter damage, 100, 172
White matter lesions, 175
White supremacy, 237, 306
Wine, 22, 125, 132, 263–266
Wisconsin, 29
Womb, the, 18, 48, 149, 165, 231–232. *See also*
    entries under Pregnancy and
    Pregnant
World Health Organization, 30, 166, 267
World Trade Center, 4–5, 6, 213

X-chromosomes, 16, 189

Yugoslavia, 28

Zygote (fertilized ovum), 91, 179